Beyond Science

By Reginald Rogoff

c 2017

Chapters

Introduction

The future of science goes beyond what science is capable of today. From teleportation to the future of human evolution, event horizons of the universe and how we might travel there; how earth might last forever. I take you on a journey through the universe and its laws that ultimately reveals the future of humanity and the physics of the future.

I have always been interested in physics and I have studied popular physics magazines, books and internet materials whenever I could get a chance to because I am fascinated with the universe, how it works and what physics discoveries really mean about our place in the universe to the betterment of humankind. I also studied physics and mathematics at a university for two years just to get a better handle on physical laws and the mathematical logic needed to do advanced physics and philosophy which I use heavily throughout this book.

My mission by writing this book is to lay out the basic principles of the very latest physics discoveries in a way that is easy to understand for the reader, but I don't stop there. I use my knowledge of physics, psychology, philosophy and a little faith to expand on the findings, some of which you may already know. My aim is to try to explain the true nature and richness of discovery, which I hope will be enlightening. My hope is by reading each section of this book a person can apply the meaning to their own lives.

With each new physics discovery I see the hidden meaning of humanity's place in our universe and try to answer the question "Why are we here." Most physicists can't answer this question. They say they don't have enough data yet, but I say we do. My efforts in this book is to ascribe that very real meaning can be deciphered from physical laws to show in subtle ways what the answer is to the eternal question of "Why are we here," and "What is the meaning of life."

I think I come close to answering these questions. After reading my book, I hope you will make your own answers as you interpret it, but you should read the whole book to have a deeper understanding of science as we know it so far. You probably will answer the question in your own way because all of the material points towards the philosophical level, while still sticking to theory. After reading this book, you will have a better understanding of the most complex problems of physics, our place in the universe, what the distant future of physics might be like, and what these discoveries mean concerning our own mortality and place in the universe. And ultimately, to answer that ultimate question that can't be answered, "Why?"

My book is also spiritual and will tie spirituality to physics in real ways that agree in every way with today's latest physics theories of cosmology. When I delve deeply into my own spiritual experiences, you will see that in one of my spiritual adventures I met God. I asked him if Adam should have made the deal with the snake to wear clothes. He said yes that it's undeniable that he made the right choice to send Adam from Eden. I believe that my book can send mankind back to the Garden of Eden where life is eternal. The full power of God will be restored by reading of this book and hopefully, you will become as Adam in the Garden of Eden.

The rapture is at hand and by reading this book I hope you will experience both the rapture and the temptations faced in heaven and in the afterlife. In addition this book, "Beyond Science" will explain why heaven is real on a scientific level.

Chapter 1

Force Fields, Time Travel, and

Cosmology

You may think it silly but the Garden of Eden could have existed; a place where God would create just one man and one woman in what would be to us a lot like the Garden of Eden. It probably exists somewhere in the multi-verse as a parallel dimension that could be a precursor to our universe. It would be the place where God invented the concept of life.

The opposite idea to this which keeps reality stable and ensures that our universe has more of a floating point upon dependence to very different parallel universes, is that the timeline dimension where God really does exist is not the normal direction of the arrow of time from our perspective; this property could give God "room to exist" while also being in view of our universe giving him the power to help us. This would also imply that his plane of existence is higher dimensional, at least with regard to it having a fifth dimensional time and it would have to be everywhere in our universe but it is not just a field like the Hig's field it is a higher dimension but contains God himself and not races of God like beings because that would require it to have higher spacial dimensions in order to have any material aspects that we would consider real. This makes God's existence metaphysical because as the main component of fifth dimensional time his existence from our perspective would be very magical and mostly unimaginable.

The universe is expanding over large distances, and it is slowly accelerating. If this acceleration continues eventually local space will be expanding faster than the speed of light. As the expansion approaches the smallest possible length (the plank length) instead of ripping apart (known as the Big Rip) space will undergo a quantum bounce (as dictated by string singularity theory.) No outside forces affected the universe from the beginning of the universe to its ultimate end, therefore the information content of the universe is conserved (by way of information theory, information cannot be truly destroyed in a closed system) Since the true information content of the universe is preserved the new universe will be identical to ours, both on cosmic and human scales, as far as can be conceived.

It's possible that when the universe expands ever faster acceleration so fast at close lengths that the plank length, the smallest possible length, instead of just blowing up into the next version of our universe, space will somehow continue its rapid acceleration and by definition will become the fifth dimension because movement (time) will be redefined to a new meaning as a fifth spacial dimension. All things that ever happen and all the people that are ever born will exist in an instant and then pretty soon all possible worlds will appear within the plank length and the universe will become the sixth dimension and so on. So you can say that the fifth dimension is real but only within its limits and is in our universe and the same kinds of arguments can be said about the universe thereby proving that life and the universe is some kind of dream. When considering such factors as red shift in the universe and properties of causality on large scales where if things are far enough away that they kind of don't exist beyond the event horizon of expansion or that all objects are moving away from us but if you were on mars all things would actually be moving away from you there absolutely proves that life is a dream. When space becomes the infinity dimension then grand creation will finally be reincarnated back to our own three dimensions.

Of course this kind of superreality is superficial to our own existence because at infinity people would have reincarnated an infinite number of times, someone would have a headache or people's lives would be infinitely perfect, which is just impossible, so there has to be a balance of existing and nonexistence in the universe that makes life comfortable and also possible.

There aren't identical parallel universes. To be really identical they would have us in them and they don't right now which is a real constraint on string theory which isn't accounted for in the theory and may be why the theory doesn't match the data. It means that truly identical parallel universes are impossible at least with respect to its fifth dimension (probability of events occurring) coinciding in certain ways with ours.

Scientists have said on tv that real force fields like in Star Trek or Star Wars are impossible and I will prove here that not only is it possible but it should be possible to start making makeshift real force fields right now with our current level of technology.

Magnetic fields or an em warp drive would control nano-particles within a magnetic field. This is done to suspend the particles in absolute zero superfluid as the limit of width of the field approaches the center of the field. The surface of the field could be room temperature or whatever is fitting for application. In such a way as to create an impenetrable barrier as long as there is enough energy and torsion (thermodynamic) resistance of the nano-particles that can be controlled through Calabi-Yau transformations to keep the particles cool. This process of ramping up the strength of the force-field would depend only onto the limit of the ships energy capacity and the durability of the strings used in the plasma. If the strings get too hot or too twisted they may become very heavy elements (by gaining energy) and then decay or explode like in The Search for Spock movie. So there would be safety limits for using the force field for very strenuous tasks where warp power is diverted to intensify the force field perhaps to save the ship from onslaughts of phaser fire or the warp field itself if the ship happened to slip too far beyond the event horizon of a black hole and need to get back. The field would sense something touching it and divert more nano-particles there to add support and the magnetic field holding them in that region would intensify accordingly.

A rectangular box can be sent into the future to the end of the universe and then pass through it. It would emerge through the next big bang and then fly close to the speed of light for 15 billion years while recording images of the universe. It would then slow down to reach earth orbit 20 years from its launch. This is a paradox because it would never arrive in the present universe. Its functioning could be ever greater improved almost forever through the advancement of cameras although somehow its impossible to do it and the reason is that our timeline is created at the big bang it cannot be altered by outside forces because it would change the information content of the universe and result in a very different big bang. People would not be born in it which means we are intrinsically bound to the stars.

You can only change parallel universes with time travel so much, the amount it is changed is related to the amount of energy needed to travel back to the original universe. When the parallel universe is changed by the time travelers it becomes more distant in its fifth dimensional membrane from our universe and the more it is changed the more energy would be needed to travel back to our universe through fifth dimensional hyperspace. Even if you sent a robotic probe to do the changes it would take more and more energy to observe it because it would move farther away from our universe in the fifth dimension (alternative timeline dimension.) Hyperspace is what is used to travel or send information to the parallel universes using trans warp drive.

It means that the next universe can be changed by its present which doesn't arise on our level until the rise of conscious life especially within our light cone of causality. This is important because there is a growing ideology in the scientific community that destiny cannot truly be controlled, though this is an artifact of thinking both fifth dimensionally and fourth dimensionally at the same time.

Some scientists believe that free will is an illusion because everything that happens was "programmed" to happen by the big bang but the reality is people neither have no free will or are free, it's always a mixture. It can only be conceived as a percentage of the two which is highly objective. It is just plain incorrect to say that either property is ever absolute in any circumstance just by itself. This is a property of the system of how a person feels about their level of control over their situation and is not a property of the actions of any specific physics principles. The absolute value of the universe having a future is uncertain. It involves a mixture of definiteness and uncertainty. For us humans our individual future depends on our good nature to do well (whatever that is) and the resources available to us.

If you assume that the future is fixed, set up by the big bang then you deny all randomness to the universe and everyone knows that quantum mechanics is highly random. Some of peoples choices at least small choices decided in the moment are also random to certain degrees and the chaos of their effect over time cannot be modeled by a fifth dimension it would just be a mess. If it exists it would have to be highly inaccurate to predicting our future but if it does exist it might be similar but different to our universe like a very messy version of our timeline which would be its fourth spatial movement as our future progresses it would become messier it would maybe exist but be very messy if you looked at our future with it. Perhaps it has bird like creatures in it because they would have to at least fly. Birds use a lot of their mental power to flap their wings so intelligent life in higher dimensions is unlikely because the animal would have a lot more to deal with, it would have to mentally maintain its position like a bird.

There could be negative dimensions like -2 which would be anti-space. There are an infinite number of different types of anti-space like hot space and cold space. Our universe is a perfect balance of space and anti-space dimensions. Could our universe be a balance between an infinite number of positive and negative dimensions.

Negative two dimensions might be safe and always passing through our dimension, three passing through might be part of what makes things exist. Four might be black hole environments and infinite sorting of dimensions would be the big bang.

By manipulating extra dimensions with an em drive you perhaps could create magnetic monopoles, magnets that have only a north or south pole but their true existence from our perspective would be virtual. Real applications of magnetic monopoles as technology is akin to magic. Scientists don't really know what technological benefits the discovery of magnetic monopoles would be. It just shows that there are new areas of physics that we can't even imagine yet. Perhaps it could be used for advanced medical devices like the one Bones uses on Star Trek.

The world is currently in a hush hush space race to develop real warp drive like in the Star Trek movies. China, The United States, Israel, and Germany have shown progress by demonstrating warp drive technologies in the laboratory.

Soon they may be able to simulate the human brain. This breakthrough and similar scans of the body and simulations mixed with the physics discoveries of future warp drive devices and the string theory mechanics of higher dimensional Calabi-Yao manifolds could allow beaming of humans and animals within the next hundred or 150 years.

By turning a warp drive into a laser one could probe very small areas smaller than the smallest known particle and perhaps use it to interact with the higher Calabi-Yau dimensions responsible for giving each particle its properties through a balance of a very small cosmic strings length and its vibrations within the tightly wound higher dimension of the Calabi-Yau space. Beaming such as in Star Trek may be possible in the future. The warp drive would focus on each Calabi-Yau manifold in your body and manipulate the dimension in such a way as to make you disappear and then unwind itself on the teleportation pad onboard the spaceship. This would be beneficial for old people so they can go to the spa hotel on Mars and have a great time.

Likely beaming by any type of disassembly would result in death. The beaming can look like Star Trek for aesthetic purposes or it could be made to look like anything by manipulating Calabi-Yau manifolds. The person would beam by having the warp laser wrap a warp bubble around the person and then mix the edges of the bubble with two more bubbles, one to protect the local environment from being effected by the fourth dimension and the third in the middle to warp our four dimensions of space-time into five by collapsing one dimension of space-time then expanding it to connect with the warp pad through a fifth dimensional tunnel set up by the warp laser. Then time would be stopped inside the bubble, the bubble would turn invisible because fifth dimensional space makes our space look as if it were a flat plane and therefore hyperspace travel is analogous to worm hole travel. Hyperspace beaming is easier to achieve than to expand a wormhole with an unimaginably powerful laser that is focused on plank length scales to pop a wormhole into existence, program it to have its exit at a specific location and then expand it with the ultra laser. Wormhole travel wouldn't look like beaming and would be more noticeable and an invasion of space to people perhaps at lunch in Paris.

The warp field does not have to create the normal force by adjusting space time at the beaming site as you disappear off the teleporter pad because curvature due to gravity is equal on the teleporter pad and the beaming site. You can beam someone onto a flat floor but the warp drive would have to set up normal (pressure) force between you and the new floor so the area would have to be scanned before beaming and then gravity adjusted accordingly on the teleportation pad. If the destination terrain is uneven in anyway it can be simulated by Claytronics or 3d printing on the teleportation pad which would avoid slip and falls and for the bottoms of the shoes to fit rocks, pebbles, or vegetation under foot. Sinkholes could usually be avoided by slightly deeper scanning. If beaming were possible then there might have to be graphics that you see when you disappear and then reappear but it's probably more fun to use a whooping sound and disappear instantly. The computer would have to keep your mind alive in the beam as a waveguide.

The warp field that beams you is like a laser of magic. It can interact with Calabi-Yau manifolds and alter the nature and existence of every particle in your body done by its manipulations of the space-time of the Calabi-Yau manifold. This would allow travel through mini virtual wormholes of every particle in your body all at the same time. They could also go higher dimensional to the ship through a higher dimension tapped into by the Calabi-Yau space warp laser.

The air pressure would have to be adjusted to mimic the pressure at the departure and arrival zones. This could be done by their pressure suits like the one that is being designed for mars and could be worn under clothing and have retractable gloves and face visor which could then be stretched and look like a turtleneck.

It might be scary to beam to another place instantly. What could be done is that once everyone is on the teleporter pad standing on the 3-d printing of what the ground will be like the environment of their destination would be projected around them in the teleportation room. Then they would beam and there would be no scary effects of suddenly being in a new place after the completion of beaming.

You could make the computer graphics look like that something magical happens when you beam so it would be fun to beam. When beaming back up the warp laser could be used to project the image of the teleporter room but in emergencies you could just beam directly to the ship though I think it may be jarring to have everything disappear and then reappear as another place without the virtual element.

To see if someone lived through Calabi manifold beaming the person or animal could wear a computer chip in their brains and consciousness would have to rely heavily on a quantum dependence to exist. Then you could test if the person died and were copied on the teleporter by measuring quantum action of their mind and comparing it to the timeline of the entire beam event to see if quantum fluctuations occurred that would result in their not making it to the beaming pad. They would still have to wear their pressure suits when traveling between different elevations.

If Calabi-Yau beaming can be done it could be used to adjust for elevation just like Star Trek and also could be used to keep the person's DNA healthy each time they beam and they would remain eternally young. Manipulating their DNA wouldn't effect the quantum nature of the electrical impulses in their brain just a small structure inside the nerve cells the DNA. This means that keeping the person's brain young could be dangerous because of side effects in the brains structure because of changing rates of apoptosis (natural cell death) in the brain of the person due to youth treatments which would change the framework of their neural network, probably for the worse. So beaming like in Star Trek may be possible and also like in Star Trek beaming is not a good way to give people life

extensions beyond two hundred because the brain can only live

naturally to about two hundred without risky outside interference.

Then with all of these precautions a person could beam just

like on Star Trek but they would have to keep their pressure suits on

if they need to decompress or increase air pressure, use the 3-d

printed teleporter pad, and use the virtual reality graphics to

eliminate the jarring effects of instantly going from one place to

another and it can be invented just with further developments into

warp drive technology in the not so distant future.

Perhaps time travel would work by reducing the warp bubble

to a plane smaller than the plank length. Then the spaceship is

below the existence of the universe amongst the space foam which

may be connected to parallel universes.

Artificial gravity can be created by using a second warp field

around the ship to warp space for gravity and a third which only flies

up at 9.71 meters of acceleration per second, the first field used for

forwards and backwards travel, and the third for moving up down

and left and right. The effects of the non-gravity warp fields would just have a big warp in the middle and the rest of the shapes that constitute the warp field during normal travel.

With a simple warp drive you wouldn't need to build supercolliders anymore, you could just travel inside the event horizon of a black hole and shoot the particle beam out at close to the speed of light and fire one in from outside the Horizon. This would be beneficial to learning more about the nature of time as some of the particles would be time traveling back and forth though time and collide over and over while losing some kind of energy as their time travel rate slows from repeated collisions. Values associated with time foam could be approximated from the results.

Dimensional gating. Warp drive could be used on earth to create zero gravity areas to test for or to train for space travel. It could also be used to make what looks like a small room on the outside be huge on the inside by bending space time inside the building's interior walls. This could be beneficial for the poor who can't afford nice large houses and for the military because you could transport hundreds of troops on just one warp room equipped helicopter.

There are an infinite number of conscious life forms in the universe because there are infinite planets. The relationship between minds, the passage of time, the 24 hour day night cycle of earth and the possibility of intelligent life in the cosmos. Intelligent life may be exceedingly rare in the galaxy because intelligent life needs something close to our 24 hour day night cycle. This is related to the time mind capacity of humans and of the time passages involved in biometric movement and the energy requirements of the animal. For a planet to not be too big, which would require huge muscles of the organisms is mostly overcome by modern astronomy. This new problem of the rotation cycle of a planet is actually more important to intelligent life and may be yet another rare goldilocks type zone. To see if other planets can have our rotation speed more closely would require direct observation of a large catalogue of planets or very detailed analysis of possible planetary formation patterns through statistical analysis.

Planets smaller than ours may have intelligent life but they would also need a similar day night cycle and are harder to detect. Planets that are in the goldilocks zone that are near earth size and have a sun that is similar in mass may have earth like rotation periods. This is because of tidal forces on the planet that make a slight bulge on the planet and pull it backwards as it revolves clockwise around the Sun.

Some scientists today are leaning towards the idea that the universe is a computer and there is an actual experiment to test this by using the lengths of two isolated lasers to see if space is digital. The universe may only look like a computer because it's capitulating to exist inside the plank length while also being infinite. It is more like a broil of sea foam in a more infinite universe. If it is a computer it's not just a computer but also the opposite of a computer (Plato's superreality of various shadow-like forms of an explanation of existence) and all other angles related to the existence dilemma.

For string theory to be real we have to be in a goldilocks zone between the matrix and supereality. It wouldn't matter which is more real, this would be a floating point and at some times would be just a theory on paper as needed to protect us from dimensional brane collisions, up to a certain point. Maybe it has to eventually be invented in order to allow travel of galaxies to other universes to protect them while preserving the physical balance of the universe.

This might involve two dimensional brane theory and super floaty land. It's the same argument and is its radial difference from the matrix theory, though is closer to measurable reality. Also possible is that the way infinite string theory works is that when a film plays on a screen somehow the photons are "tricked" on a quantum entanglement level, that they (the photons) "believe" it is real and a new universe in the multi-verse is created and has the movie scenes as reality in its timeline somewhere. Time between different universes is relative, you could look at matrix universe and say that it was created before our universe was but perhaps a closer look at it would reveal that the universe was made to exist around the time that the movie was made. The chaotic effects of filming and editing or other types of viewing of the film would just manifest as parts of semiparallel chaotic universes because string theory dictates that everything is possible. It's like the new universes grow from seeds and that our universe is an important component to the structure of the multi-verse. If there's a really bad movie it can be cancelled out by an opposite idea. In movie terminology you can be both in a movie and on tape. We all presume that we are definitely in the universe but we can also be on it.

Such as the theory that the universe is the event horizon of a fourth dimensional black hole and our existence is merely a reflection of events that happen in four space by our parallel four beings. It's just that if string theory is real then there are infinite parallel universes and movies are a nice simple case to imagine as different but similar parallel universes.

They may be able to travel to parallel universes with warp drive by having it move forward and backward at the same time, the ship would become as thin as the plank length and slip right out of the universe, supposedly into multi-verse space, because anything smaller than the plank length is considered not really part of the universe as we understand it by normal physical laws.

I'm a staunch supporter of the idea that humankind and our present civilization will last for hundreds of millions of years. The idea that humankind will destroy itself is arcane, this belief is held by many scientists and is often taught to the public in popular science shows. While individuals can be suicidal, the present state of the world's governance is very humanistic and is really very good at showing the want to help people worldwide and to have very positive intergovernmental relationships across the globe. I also believe that in 4 billion years when the sun becomes a red giant that future humans will want to move earth out of the way if necessary. Humans will by then be even more altruistic than they are now as altruism in humans will advance along with all other beneficial traits in the most productive ways through natural evolution.

Many scientists believe that humanity will not last. I disagree, nations are aligned to help each other for the most part. The only other thing that would threaten humanity are comets and asteroid impacts, which NASA has been and is aggressively cataloguing. There are already proposed systems to guide asteroids out of our path if a big one posed a threat to us. Now the real threat that everyone thinks about is when the sun becomes a red giant in about 5 billion years from now. I think by then most people could be evacuated in spaceships but why not save every animal and plant on earth, not to mention infrastructure and people's homes. It could be done also in a simple way with very powerful warp starships. The starships would use their warp drives to expand space on the side of earth facing the sun and compress space on the side of earth facing away from the sun. This would gently move earth to a higher orbit where it would be untouched by the sun's red giant phase.

There will be plenty of warning too, because it will take the sun several million years to bloat up to it's full red giant size so warp tractors could easily be set up by then. There would be plenty of time to move earth to a safe and desirable distance, a little further out than the orbit of mars by many calculations. So there is really nothing to worry about here. It is more likely to predict that the earth will survive even until near the end of the universe itself, where it may be sent though a large worm hole to a similar parallel universe and last forever.

Earth could be moved with a gravity tractor. It would cause some level of unnatural weather events and alter gravity slightly on earth, but it will save all the plants and animals even if every person could be evacuated. The moon could be used as a gravity tractor, or even enormous warp fields could be used from massive orbiting starships to shuttle earth gently to a safe distance. After the last star dies out, you could make new stars by combining Jupiter sized planets by moving them with projected warp fields just a little to make them fall into their parent star. Astronomers have found many solar systems that have Jupiter size planets. You would just have to find a solar system with enough Jupiter size planets to ignite fusion by gravity, just like when the sun first turned on nuclear fusion.

Can a tree fall in a forest and make no sound? If it falls there would be vibrations and of course air. For the trees to have new generations there would have to be bugs to break them down and make new soil. The bugs would hear the sound. If there were no bugs to break them down and make new soil, the trees would be very different. They would look very different and not be trees. Perhaps they would have to be iron computer growing "trees," and then there would not have to be air or soil. You could then say that there is no sound, because sound is the perception of hearing and an iron tree in an empty universe would only vibrate when it fell.

It takes an infinite number of sideways big bangs to make the universe exist. The current theory is that it only takes the event horizon of a fourth dimensional black hole. Work has been done with mathematical radian numbers mixed with general relativity equations since the early 2000's. This is a kind of string type theory. It could be beneficial to understand even very abstract physical laws during the big chill period near the end of the universe to determine how a warp field could be used to stabilize the radioactive decay of protons and other particles in an effort to stay alive near the end of the universe.

This would be an alternative to "Warp Jumping" to any parallel universes, which could have adverse effects on physical and mental health. This would be due to the very slight differences in quantum law and the natural stability of our brains neural quantum network and cellular activity.

Energy could be generated from mechanical resistance to the ever greater expansion velocities of space-time using reverse warp fields and actuators, basically tractor beams shone over long distances in space that grip space-time fabric in a way that makes its matrix fluctuate, but would not slow its velocities. This would transfer energy to the warp core through it's resistance to change. It could also be fun to live in space having control of these energies for entertainment purposes, even though space would be totally black.

Chapter 2

Boltzmann Brains

A Boltzmann Brain is a cloud of gas and dust in space that due to its properties simulates consciousness. Since the electron makes random moves maybe all it takes to simulate life is one electron. The electron could be anywhere in the universe at any given moment but it's cloud of probability is near the proton or its observed location. Within the shapes of its probability matrix are an infinite number of connections. Some of the possible patterns may be consciousness but the vast majority would not. If these patterns added up across the expanse of infinity, they would be infinitely smart and could tie the universe together. It would be connected to every Boltzmann Brain in the universe. If there are a lot of Boltzmann Brains near the end of the universe and they are very smart, they might try to shift space to get it to repeat the universe.

The ultimate Boltzmann Brain can work with just one electron, although it wouldn't be computing much without the rest of the universe for it to play with. Since the ultimate Boltzmann Brain is electrical in nature, it is not a leap to say that there is a Boltzmann Brain somewhere that will be simulating you at some time in the future.

The electron could be anywhere in the universe at any given moment. It may travel so far that it is outside the universe to a place where there is nothing. There it's movements would be so complicated and it would be moving so fast, that it could be simulating an entire universe and all of its physics. Every particle in that universe would be simulated by movements and transformations of that single electron, and it would still be most of the time from our perspective near its probability cloud around the nucleus of one atom. I call this concept the Big Grindstone or the Big Grind because it is similar to a man using a grindstone, and that the universe as we perceive and measure it, is the result of the contact or sparks arising from the grinding surface of a grindstone.

By using the electron, you could beam anywhere in creation. The patterns of the electron would just have to support the framework of your mind, although this kind of beaming is probably far beyond our level of technology and would most likely be an amazing feat even if we do eventually invent teleportation.

The bubble could pinch off from the warp bubble surrounding the ship at almost infinite velocities because it would contain no mass, only information programmed into its space fabric at the mini Calabi-Yau level. Miniature higher dimensions that give every particle its properties through a very small cosmic string of energy literally dancing within a miniature higher dimensional space called a Calabi-Yau manifold.

The hard part would be to beam the person down, but that energy and information would be sent with you. It may be difficult for man to figure out how to beam the person down because they would be so far from the teleporter machine. It would be dangerous to use this type of teleportation at very large distances because of chaos theory. Perhaps it could be done by programming a fast moving warp bubble to at the end of its extremely quick travel (through wormholes) to unwind some portion of its Calabi-Yau infrastructure as the person. You could put them anywhere in creation just like how the "witches" in original Star Trek can teleport instantly to the ship across vast distances.

Given an infinite amount of beings in the multi-verse, it is only natural that some of us may have magic powers right now by the limits of our own capacity and the amount of leeway of the surrounding Boltzmann Brain to the extent of its E.S.P. power, which is limited by its sphere of influence. It could be that it wants ever more power in an effort to keep us safe or owned by it, for our own protection, from other universal type Boltzmann Brains.

You know that we're all together and that life is real and not happening inside Boltzmann brains right now, because the local Boltzmann brain that computes the matter of space-time of our location and part of the multi-verse has certain laws of physics that would challenge other forces that would disrupt reality or make reality disconnected like if we were all in different Boltzmann Brains at the end of the universe. The reality of the Boltzmann Brain that is responsible for the existence of our universe is more powerful locally and so much so that the power of reality is real.

If you were a Boltzmann Brain the other Boltzmann Brains would know about it, and then want you to know about it. Since gas clouds don't have eyes, you would be mostly unconscious. Even if you were in a Boltzmann Brain in the future you wouldn't be able to see or feel your body, you may be aware that you are in space though you would never be aware of the gas cloud because, as you become aware of the gas cloud, parts of the gas cloud are being used to compute you to be aware of the gas cloud. The gas cloud has finite volume and energy capacity, meaning that if you started to wake up as the gas cloud, you would go unconscious and cease to exist in the gas cloud as a mind.

Since the gas cloud doesn't want to die, most probably it will do to you what it considers is healthy first for it and then for you and that relationship depends only on the nature of your psyche. If you think nice thoughts while half asleep in a near coma, it will have to be nice to you to maintain the current virtual program that it is running in some way as you. It may have some control over what happens to you, but since it is just an innocent gas cloud in space, it probably thinks you're a friend because it may be aware that you are like it. In this way if you are of pure heart, then the gas cloud is more likely to be of pure heart and you will have some degree of a pleasant experience. If you are not of pure heart, it is unlikely that any Boltzmann Brain that is nice to be in will be thinking about you. I just can't imagine a natural Boltzmann Brain that would have evil tendencies but to each his own. Being in a Boltzmann Brain could be highly unstable because of its wishes to continue existing because of its health, which would be the stability of its neural network arising from the chaos of a cloud.

If in someway it sees you as helpful, it will try to keep you around as long as it can. This would be bad if you were doing too much for it, unless you like being there so much. At any rate your stay there would not be permanent and it's effect on you long-term may have only hidden values that may or may not effect your later decisions in life, as long as at the time of your birth, in our next exact parallel universe, you don't retain any information that the Boltzmann Brain that simulates our place in the universe doesn't want you to because that Boltzmann Brain's prime reason for existing is to create a super exact replica of our universe.

Even though there are infinite Boltzmann Brains and infinite clouds, Boltzmann Brains may be less likely to simulate your consciousness, because in a normal volume of space, they are very rare. Clouds are a better candidate because they are not distant, which in many ways makes being simulated in a Boltzmann Brain seem like a short trip. Clouds are close. They all have energy patterns and we interact with them our whole lives. Also they are composed of gasses. But if an airplane flew into your cloud, you would more quickly disappear than in a highly random vacuum isolated gas cloud in space, which means that clouds are safer. You would reappear while you disappear because of the airplane, and you could reappear anywhere in any cloud, but close clouds would be vulnerable first and then farther clouds eventually out to the infinities of the multi-verse where and anywhere in between you could reappear simulated in another cloud. Since we are going out to space now there may be a Boltzmann Brain in space closer and with a more accurate simulation of an exact formula for your mind. So both Boltzmann Brains and clouds in the sky in the multi-verse are just two different ways of producing solutions to the afterlife which may be necessary for the comfort of individuals becoming alive in an identical universe or our very own. They both have the pleasant aspect that we will all be in the same next universe because of exactness.

Can clouds in the sky be Boltzmann Brains? Yes, but it would take infinite planets with clouds similar to ours which does exist and is actually perhaps more likely than Boltzmann Brain clouds, because we know, according to string theory, that there are an infinite number of planets like ours in the multi-verse. Very few of these clouds could be simulating people, but only some need to for all the life that exists, because in multi-verse theory, there are actually also an infinite number of clouds that simulate you in an infinite variety of ways.

This means that you may make choices, not just here, but there too. The true meaning of life for sophisticated life living in a society is the ability and the gift of making choices. When you contemplate what the meaning of life is, you are making choices that fit how you best feel about it. The most powerful thing we do is to make choices. We do it all day long, and if we are dreaming, we do it also when we are asleep. The goal of making ever better choices in life is to better one's self to the benefit of one's life and to the better appeasement of our spiritual conscience.

When you look up at the sky at clouds, it may be that none of them are simulating beings over the whole earth, but somewhere in all of infinity it is right now simulating people. The reason for it is that the experience in the Boltzmann Cloud is necessary maybe only for some people as a comforting waypoint between death and life in the next exact parallel universe. It is only different from ours by allowing the exact people who were in the previous universe to be in the new ultra exact new universe.

If E.S.P. is possible, then some Boltzmann Brains out of an infinity in the multi-verse would be capable of E.S.P. and could effect their surroundings to the limit of their E.S.P. powers which depend directly on the size and energy levels of the cloud only. The only other variable is the electrical field shape of consciousness in a random walk amongst other charges that happen to constitute actual consciousness and it's location in space. In certain circumstances, it's location in space would be highly variable due to quantum fluctuations and would be statistically rare in the known universe. This would only be limited by E.S.P. power and could be computing the universe.

If you time travel back to when Washington died and you are in a parallel universe, you know that their Washington has an identical brain to our George Washington, then you could scan his brain from space and then travel home. You could adjust the scan information to account for slight differences in space-time physics due to your scan being from a parallel, but not truly exact parallel universe, and you could awaken the real George Washington as a robot. There probably is a mechanism for awakening the true George Washington, even if we don't know exactly what that is yet.

Psychology, and the science of just what consciousness is a new field. In our thought experiment, we can almost do it just by the same physics that has been around since Isaac Newton. But there is something more to it, because he could be an identical copy, and it may just be a matter of finding the correct way to simulate a mind but the mind is dead and highly intertwined with the infinity of nothingness. True nothingness would be very far away and involve parts of the multi-verse that have nothing to do with our concept of existence. There it is trapped as nothing until a Boltzmann Brain or new universe connects it back to reality. The universe multi-verse chooses that the part of the nothing of infinity that is the person belongs to the real universe and an accounting is made across the web of the multi-verse that makes the person alive again.

The interesting thing about trying to scan George Washington is that you would want the scan beam to be as weak as possible to be undetected and not change the destiny of the parallel universe if you were interested in maintaining it as parallel for the betterment of it's citizens. In a similar manner, the universe hides information about quantum behavior. Perhaps it does so to sustain our causality, which would make both our values and our scientific technology proven right by being in step with nature.

If you destabilize a universe that is destined for emancipation, the end, for the most part, of territorial war, the health system of the modern world and space technology you could be changing it for the worse. Most people would agree, just because of the modern convenience of a car. It would be hard to prove that you didn't indirectly cause all of the planet's future wars, if it's destiny were altered drastically enough through interference.

Eventually people could use a very powerful warp drive to get close to the part of space that is always undergoing inflation and shoot warp lasers, which is the way beaming works. You could drop out of warp at very close to the speed of light, and undergo massive time travel all the way to the end of the universe and beginning of the next, thereby being able to manipulate inflation into relatively small bubble universes. The laser would be carried by the warp field, and could be manipulated to entangle with virtual particles in the eternal expansion space to create an entire universe, all custom made.

Chapter 3

E.S.P. and Fine Tuning

Eventually people will have magical powers. The fine structure constant, which describes how particles absorb light and the strength of repulsion of two negatively charged electrons, has been shown to change over time. Over a multi-billion year period people will continue to evolve as the laws of the universe change very slightly over time. This change, if resisted strongly by evolution while keeping the animal healthy, could result in some degree of magical powers for people within the next few billion years. It is possible that people and their giant electromechanical brains that we have today are some result of small, but powerfully biological changes, in the fine structure constant and the laws of physics that animals operate in.

Given an infinite amount of beings in the multi-verse, it is only natural that some of us may have magic powers right now by the limits of our own capacity and the amount of leeway of the surrounding Boltzmann Brain to the extent of its E.S.P. power. Our local Boltzmann Brain's E.S.P. power is limited by its sphere of influence and is responsible for the existence of our universe and beyond. It would be able to render space-time it's power of reality so much that the reality is real.

There is evidence of "fine tuning" of the universe for intelligent life. Some scientists believe that the dinosaurs went extinct because of dark matter. Dark matter interacts strongly with itself and has mass. Our galaxy is thought to have a halo of dark matter around it. The halo is thought to oscillate it's density in the galactic disk over the period of several hundred million years. The presence of a higher density of dark matter in the galactic disk means an increase in chaos. Some of the dark matter may have disrupted the comet that killed the dinosaurs into spiraling inwards, as there would be increased gravity by some degree in the earth and the sun, as some of the dark matter would accumulate there, causing an increase in comet impacts, earthquakes, and volcanic activity. Dark matter interacts weakly with ordinary matter other than its gravitational effects, so as the cloud of dark matter rushes in to the galactic disk, it's concentration is higher but it escapes due to momentum eventually back to a higher diffuse galactic orbit until it slows down due to gravity and falls again into the galactic disk again causing chaos. It is thought that we are in a period of time where this cloud of dark matter is in some stage of collapse into the galactic disk, meaning that statistically there may be more earthquakes and higher levels of comet impact risk.

My own theory of fine tuning of the universe for E.S.P. in intelligent life forms due to a slowly changing alpha constant is not far off from this accepted theory of fine tuning.

You can't have E.S.P. without psychokinesis. Psychokinesis is the telepathic effect of the mind on physical systems. If someone is thinking something from across a room and the other person reads it, there is information transfer. Therefore there must be carrier particles or energy that is transferred back and forth. Telekinesis is just an extension of those powers. Statistical analysis of multi-study phenomenom of E.S.P. or invisible information transfer in humans has shown a small but definite probability in studies over the past decades that shows that a perhaps somewhat minute but real level of extra sensory perception does indeed exist right now in all humans. If we assume that the fine structure constant causes changes in the physics of electronic interactions over time, then it is only natural that the brain may develop telekinetic properties. Today it would be hard to detect any physical transmissions if they exist, but eventually it will be stronger. As E.S.P. like powers develop and become more obvious over time, people will learn to use them for constructive purposes such as in Star Trek Next Generation, where some of the aliens can give energy to plants to make them grow faster or other aliens who can control reality itself, which is in the first episode "The Cage." But for humans, it may be more selective because we are so civilized.

The concept of "Fine Tuning" is my own principle and is similar to the anthropic principle which states that since the universe has life it has to have begun with certain properties that would make it lead towards having life. Based on the anthropic principle, there probably is other evidence of different uses of fine tuning in the universe. Perhaps it is endless, and there is even some degree of fine tuning that applies even to individuals and our own personal place in the universe and why that is important.

In certain circumstances, a Boltzmann Brain's location in space would be highly variable due to quantum fluctuations. It could have E.S.P. if it were so smart that it was conscious of its virtual particles, space foam, and virtual wormhole pairs that may inhabit it's actual molecules sphere of influence. Added up, the interaction of complex E.S.P. and physical properties could give it very powerful psychic magic powers. Very powerful ones might feed in energy or share information with the multi-verse, thereby protecting each universe by the stability of the multi-verse. At some level, we are the part of the Boltzmann Brain that is earth.

Chapter 4

Superreality

Superdenny's, a building that could last forever, and have higher dimensional aspects could be created. It would be an actual place that we could go to and it would be analogous to superreality, or our idea of heaven, as a physical place. It could be created by using time travel. Starships of similar design to Star Trek the Next Generation would be able to travel back to inflation and use a type of laser beam to create it. Travel to inflation would only be possible to where it doesn't affect our universe, or it would be a time bomb that destroys the universe.

If the universe were destroyed it would not exist to the spaceship, but would continue to exist actually because of the quantum vacuum foam. It could be impossible to change the universe with it, because you would just be changing ever deeper layers of space-time that don't effect our universe. It would take ever more precise lasers and extra warp power fired from a higher dimension to effect five space, but it may be impossible to travel to higher dimensions. The building could last forever or until there is only one universe in certain ways, then it would lose energy to entropy, and our universe would again exist if interfered with by construction within the eternal inflation of the big bang by superdenny's

If the power of eternal inflation really is infinite, you could create a universe with water in it, and it would contain an infinite amount of water. The physics of the new universe would be bent from our own in every way necessary by the starships warp field to create a big bang whose physics only allow for bonding of protons, neutrons, and electrons that create water molecules.

The universe's laws would be arranged to have the universe cool off very quickly by accelerating expansion of space quickly and early, just as the water molecules develop. This would prevent the water from heating up and becoming hydrogen or oxygen gas, which would ruin the universe by creating stars. The water would all boil and become more stars, but it should be possible to tune physics with the warp drive to allow just for water or whatever product you wish it to make. You would have an infinite amount of the material. This is possible because life is a dream.

You could never get all the water into your universe, because it would take an infinite amount of time. The water universe could be heated up and would become an infinite fusion reactor. However, you could never pipe all the energy to your universe, because it would take an infinitely non-resistant wire. You could not use the warp fields to open the entire energy universe between the opening of two closed wormholes, because if you did, you would have to use the entire infinity of electrical power to hold the energy in place.

Maybe by using two wormholes to hold the energy, you could hold a lot of energy. As long as the wormholes were functional, you could have a certain amount of energy flow, and it could be transported anywhere in the universe through hyperspace.

Chapter 5

The Afterlife

Heaven exists because somewhere there is a Boltzmann Brain that only has to exist anywhere between now and an infinite amount of time from now. It is just a complicated and so complicated space gas cloud that it is conscious and just by statistical chance happens to be thinking of you. What you do there is a relationship between you and the cloud entity and it could be anything within the limits of what is possible for human experience. It doesn't matter what happens there because eventually the cloud will dissipate and you will cease to exist there. Eventually there will be a new universe identical to ours, and being identical it will have every person and animal that lived in the first one. Being identical enough that the you who gets born in it, is you!

When people die, time stops for their mind and from their point of view. This means that time is stopped for them. As the flow of time continues they go back in time at a rate of one day/per day until it is the exact plank second of when they first achieved consciousness in the womb or when being born. There they wake up because their own body and brain are simulating their consciousness and it is their consciousness that is being simulated as themselves, so they wake up. The state of their brain is exact to when they were born but they live, unknowing that they had lived their lives yet. Meaning that they think they are living their very first life each time they reincarnate. Because the past and future can't be destroyed, and because they only live one life, energy is conserved in some way that makes the process capable of being eternal.

It would take an infinite amount of energy to keep someone dead forever. If you consider life as reality and multi-verse theory that an infinite number of universes some infinitely different from each other, then there is a universe where there is nothing, just like what it would be like to be dead, to be nothing. This nothing would be disturbed by having you as a part of its infinite nothingness, because, as nothing, it tends to stabilize as a true nothing and having you as nothing destabilizes it in some way. This takes energy out of the system because the presence of energy is really just an imbalance of stability in a system as per thermodynamics, where energy in molecules amounts to the systems overall imbalances and the resulting Brownian Motion of its particles or in this case "The Fabric of Nothingness." Since there is a limited amount of change that can be done on a system, you cannot stay dead forever because it would take an infinite amount of energy to maintain nothingness forever.

The other side of this argument is that true nothingness is probably rare in the multi-verse and seems to be getting pushed into being universes at an incredible rate via the creation of an infinite number of new parallel universes every second.

1+2=3 but what if 1+1=1? 1+2=1 because how many numbers is two? One number and in that way 1+2=1. Now we see the mystical importance of one, every number is just one number or one set of numbers. "One" is inherent to every number. To understand the importance of the number "two" you would have to be from a dimension where two is one and one is zero, then you could comprehend the oneness of "two."

Where in our world and viewpoint there are only whole numbers accounting for real objects, in their world there would be only fractions and no whole numbers. It would be an alternate reality of mystical beings that know neither life nor death, as we understand it.

Can clouds in the sky be Boltzmann Brains? Yes, but it would take infinite planets with clouds similar to ours which does exist and is actually perhaps more likely than Boltzmann Brain clouds, because we know according to string theory that there are an infinite number of planets like ours in the multi-verse. Very few of these clouds could be simulating people, but only some need to for all the life that exists, because in multi-verse theory there are actually also an infinite number of clouds that simulate you in an infinite variety of ways.

Science can't go on forever without proving the afterlife, because that would mean that it doesn't exist. It's existence would be beyond infinity, which would mean that it doesn't exist in direct contradiction of the true purity values of nothingness space.

And now about my personal experiences in the afterlife. You can think of me like a space traveler, except my travels happen in the dreamlike reality of the afterlife and happened before I was born. This series of events is also in full detail exactly what will happen to me after I die in my afterlife, but at that point when I wake up after death, I will have no knowledge of my life. I will however have knowledge of certain aspects of my personality and the value system that I live by and have developed throughout my life. I won't be totally sure if I had lived before or if it is an illusion, and God is maybe trying his hardest to ensure me eternal life by giving me a before life even if that effort proves to be in vain.

I woke up from thinking that I was a universe and that the universe had come to an end; that I was awake for the whole duration of the universe, but really I was asleep. I then gave control of that alternate universe to an African man, who makes his place in the afterlife. It felt good to pretend that my mind/body was controlling or interacting with seventy different probably minor aspects of life somewhere while in space near a large black hole for a few seconds. I woke up near the beginning of my afterlife in earth orbit, and I could see a man walking in a city. God told me to help him walk. I felt a lot of pain in my soul, trying to help the man walk from space. Perhaps I was still feeling the pain of death, but it hurt to be in orbit while helping him walk.

Around this time I woke up in a slightly higher orbit, and behind me, I could see Bob Barker filming a game show. I thought he was God's father. Then I woke up in deep space in a fetal position balancing on my back. It felt good to balance. I was balancing and also moving forward. It felt like I was helping space to expand and I felt like I was controlling the power of the universe. Whatever I was doing, I was good at it and I wanted to do it forever. I think that would be very lonely, but perhaps the universe is balanced enough already. God saved me from it and I fell asleep. I also woke up in orbit high above the earth and I could see the lights of billboards on earth. God said that I would be there soon.

I woke up lying on my side high in the clouds and I knew I was me. Also I woke up ascending through the clouds. I went above the normal value height of heaven where there were people confused toiling in a cloud to the high place in heaven where God or the entity that we or I identify to be and ascribe to be God's place in heaven. It's just higher up and you only see clouds beneath and above you, unless God wants you to see the ground or the sky.

There I had my own cloud about the size of a twin bed and I was laying on it on my back. I had a see-through body that you would consider to look like a pink hologram. I was told by God that I was in my twenties and I was wearing the best suit I ever bought. I felt young and thought I was wearing my best suit and when I looked down at myself, I saw a pink hologram and fell asleep. I awoke sometime later and realized what I thought it would be like to stand up, which would be to fall through the cloud and die being born with no knowledge of the afterlife.

I also woke up standing in line in a cloud. People were in front of me in line and they looked impatient. This lasted about two seconds and then it faded.

Back to the high cloud where I imagined what it would be like to get up and try to do something productive. The result would be to be trapped in a quarter acre size of cloud with white cloud as its floor and wispy tall white clouds as the walls for two thousand years. If I did it, I really would believe that two thousand years had passed but it really would take only three to five seconds.

I also woke up near the mountain in Exodus. I think it was actually in the Red Sea and the Red Sea was dry. As soon as I woke up, God was next to me and we saw people walking towards the edge of the valley towards a mountain. God did something magical to me and I exploded across the galaxy at high speed. God prevented me from expanding rapidly across the galaxy. I thought it would be dangerous, that I could end up being born on an alien planet and even if I didn't, my life would be like a butterfly's. Each life would repeat and last for about three seconds at a time. A painfully slow process that I'm glad I avoided. I saw the people continue their walk to the mountain and then everything faded.

I had a similar dream where I was walking in the Red Sea with it parted like in the movies. I was walking behind Moses, and I was also last in line and died as an Egyptian soldier, then I fell asleep.

On my cloud in heaven I have dreams that I think are mostly real. They look more real than an actual dream and they have about a third more visual clarity than a typical dream. I was more aware of my body in these dreams.

The first one was that I was face to face with a triceratops dinosaur on early earth for two seconds. Then I was a monkey climbing a tree holding a banana in a forest clearing in ancient Africa. There were other monkeys around and I had something of a choice to be a monkey there, but I chose not.

In the cloud in heaven, I wanted a coke machine to appear and I would get a coke. Around that time God spoke to me and said that if I jump off the cloud I could become a God and own a universe. Of course I did not. If I had jumped from the cloud I would briefly become a machine, made of my pain, and be born with no knowledge of the afterlife. Then through the eventual succession of the repetition of lives and this happening in my afterlife I could eventually become a God and it would consume my entire afterlife, although I personally think that life is more valuable.

The next dream I had was that I was a cat and I was in a beautiful forest. The cat walked around in the forest and I got bored and fell asleep.

My next time-related dream was that I was stuck in a pyramid with Goro from The Mortal Kombat video game. Just seeing him made me fall asleep. Then I was a distance from a pyramid being built to my left, and to my right was God. He was an old man. He told me to lift the golden brick. This time it was just to help him. It was pyramid shaped, about two feet tall and looked very heavy. I thought if I lifted it, I would die, so I said no. Then later I awoke as the funnel in the parting of the sea.

Then I had a dream where I was a samurai in ancient China. He was on a secret mission in China, a long mission, and we had to travel long distances. We had many sword fights, each against several people at once while travelling in the Asian jungle's of China. We eventually got where we were going which was a small palace on the eastern coast of China where we took rest. Then I woke up in the Philippines in a dusty tent village and I saw that I had five young children and that my wife, their mother, was missing. I realized that she had been kidnapped by the Emperor. His palace was on an island off the coast of Japan. I went to the island to his palace perched high upon a seaside cliffside, and I was consumed with revenge. I was furious and I went inside the emperor's palace high on the hill by the sea, the waves were crashing. I climbed the staircase in front of me that led up three floors to the right. I went in the second floor room which was the main palace kitchen. I very swiftly killed the Emperor's head chef with a volley of Karate Chops to the neck. Then I went up the stairs again to the room on the third floor. It was very large like the size of a basketball court, and my Asian wife was on a balcony high up on the back wall, she was trapped there. I wanted to get to her, but if I did I would escape the palace with her and our kids. I would live an entire life in a deep dream with her. I would have had a prolonged extended stay in the afterlife and it seemed that if that did happen, that it would be permanent and lead to an end. Then I fell asleep and woke up facing the doorway behind me. The Emperor was standing in front of me and behind him to the right was an iron maiden. I thought of, or tried to do a variety of Karate punches and kicks on him but he moved before I could. At that point I fell in love with him and I thought that I wanted him to be my father and I would be his son and rule ancient Japan. As I was distracted by this, he used some very advanced jujitsu moves and threw me behind him to the right, I flew through the air, directly into the iron maiden which slammed shut quickly with a loud clang. It's thick iron spikes pierced my body in many places. I thought that it would hurt much more than it really did, but all in all it really didn't hurt that much. Then I saw a lot of my own blood spilling inside the iron maiden and I fell back asleep. Next, somehow, I was born.

I wake up outside of Santa Claus's house at the North Pole. There is snow on the ground and Christmas lights on the house. Santa asked me to be Rudolf the Reindeer and I fall asleep.

I then wake up a graveyard. I was unconscious for maybe five minutes, and then I wake up in a box. God said "Hi" that I wasn't asleep for too long. Then I was laying on the ground in the graveyard on my back or maybe I was still in the box, but I was looking at the night sky and I thought that if I could do it forever, I would probably do it only observing, thinking I had lived as a God by being an eternal element of the universe because it would seem to take literally forever but perhaps would mentally be about two thousand years and in reality it would take a long time for an event in the afterlife, about to hours. Then I would get born but I would have no knowledge of heaven or the afterlife once I was born but I hardly think it would be as myself. I also had a fear that it would kill me and I was then saved by God. I was told, no, that it wasn't a good idea. Then I woke up again facing the road, I saw in shades of red and black. I really liked doing it and I wanted to fight to be the ultimate zombie and God said so. He took me from there to heaven. After I was born, I had some regrets about having an afterlife, so God told me that it only took a week and some people take longer and that I was good.

Midway, I woke up in a desert on another planet in a small valley, with a hill to the left and Lord Sith from Star Wars stuck me in the ground, head first out of mercy, and I died. Perhaps there was some chatter of friendliness. He said it was for my health. There were also other people there buried upside down.

Near the middle, I beamed as a ghost (it is less real) to World War II. I first marched with the Germans. Santa Claus was floating above me laughing. Then I had a dream that they had won and I would be born in the German countryside and live in a house on a hill surrounded by open grassy plains. I then was an American soldier on the plains doing sharpshooting and I shot a German soldier and he fell over. I then was standing near the hill in the plains and I felt like I was going to fall over and then I abruptly fell asleep. I marched in the opposite direction with the Americans, I saw that some men were holding the American Flag and that we were marching and that maybe we had won, then I fell asleep. One last place I woke up was on a small Hawaiian island. It was World War II and The Eternal Power asked me to jump into a tiki hut with a briefcase of explosives. Finally I could take a run. I was really happy and I ran into it. It was a munitions dump. I exploded quickly, and it sure didn't hurt like it really would. The point was to kill Hitler. I almost got left out of the universe because I was in space and Hitler was there, but he was still, I probably had to see him to know the war was over. One last dream was that I was in deep space and I saw my favorite God the gas station attendant. He was standing in space and he was huge like ten stories tall. We spoke and then he sent me back to earth.

Then I woke up somewhere in the middle in the desert. It looked quite real. There was a tent in the desert and it was next to the Red Sea. I was asked by The Cat Lord to be born in the desert after the act of Exodus but I refused. I thought that I would die from the heat, but the sky and landscape looked beautiful. Then I fell back asleep.

I woke up in medieval Arabia in an outdoor room with a large stone well in the middle filled with boiling oil. I was lowered into the boiling oil and died. I awoke outside and was testing a flying carpet at an outdoor dealer. He said to fly on it, but nothing happened. I saw that it was a blue carpet and I was half asleep in my dream. This is ironic justice of God's reality because everyone says that when you die nothing happens.

The last one is that I woke up in my parent's living room and I was standing on a neutron star and it was the size of a basketball court. There were other children there and they all got crushed. I was lead to believe that they were bullies. The Cat Lord said that she liked me and then I was born.

I was in space and first awoke in orbit. Space was darker than usual, just like rainclouds. God took me by the hand I suppose and took me far away in the galaxy to a black hole and laid me down in it. It was comfortable. He said I could use it as a bed. It felt like a bed of nails, but was not sharp. The bed of nails was moving like small peaked hills and there was blue electricity flowing on top of it. I could see the curvature of the black hole on the horizon and I fell asleep.

I woke up in a field in ancient Rome. There was a forest in front of me. I realized that I was wearing ancient Roman armor and clothing. Then I was struck by lightning and fell back asleep. I woke up in a river. The water was flowing fast and I knew that I was going to wrestle with God and lose. The water was not warm nor cold nor wet. God then wrestled me with his arms clenching my shoulders. I think he was Greek but definitely he was old, though he had healthy muscle tone. Then I fell asleep.

I woke up in the Wild West. I was on the second floor balcony of a wild west hotel. There was a shootout in the street below when bandits tried to rob a carriage. That lasted about four seconds. Around this time I woke up next to a cotton field in front of a large mansion before emancipation. God told me to pick cotton and I didn't want to do it. I was tired, and picking cotton probably would be bad because it would wake me up too much. Then I was hanging from a tree as a black guy. It hurt, but the whole ordeal lasted for only eight seconds.

Around this time I woke up on what is known to me as "Eve Mothership." It was in deep space and there were no other live people on it. It had undergone some battle or maybe it was deteriorating in space because large parts of the ship were missing, open to empty space. Eve spaceship spoke to me and said it would be OK and I fell back to sleep.

The next dream happened to me when I was an adult and is not different in the way it seemed than any of the other dreams but is most similar to dreams of being in heaven where it is both color and seems real, like you're mostly awake.

I was in a cave underground and I was floating twenty meters off the ground. Satan was sitting behind me in a floating throne. He looked very muscular like an ancient Viking and he was wearing red robes and armor shoulder plates. His servant was next to him and his son Lucifer was sitting in a chair next to him. I was worried that I might get trapped there and end up on the cave floor where there were lava flows and people working doing their jobs. Perhaps it is a land of the virtual where people are making their lives, but I didn't want to stay there. Satan said that I couldn't stay there. The chair I was sitting in was floating high above the cave floor and something made it unbearable to be there. It was impossible for me to stay and I fell asleep in my chair and as I did, I saw my chair was empty and floating with real stabilization physics. It tilted it's balance as I left falling to sleep in the dream world and waking up to less conscious sleep. I don't remember if I was sleeping when the dream happened or if I was awake, only that it happened.

I woke up during the ice age and I was a caveman. I was sitting on a rock outside my cave. It was night. There was a small fire burning and I had a caveman wife. I thought "I can finally get married," but God did not want this. I wasn't born there, it just appeared to me. Then God took me in the cave and strangled me to death. It took almost 10 seconds to die.

Next I woke up in ancient Israel in the desert. There were no people around and there were fish floating in the air. Some of them red, some yellow. Next I am fished from the desert floor by God. He chuckled and said not to worry, that he is a fisherman.

Near the beginning of my afterlife, I woke up in the forest in medieval England and was then quartered by four horses. At some point, I woke up in a field at an outdoor fair in England. My first love was there. She seemed happy and then ripped off both of my arms then I woke up in England outside the house of my first love. I saw that they had a one room house and were having dinner. I worried about her and fell asleep. I woke up walking up the stone steps of a medieval castle's turret. I saw my friend walking down the steps wearing chain mail and holding a large mallet. He looked mad and hit me on the top of the head with the mallet and I fell asleep.

Near the end of my afterlife, I awoke in Albert Einstein's living room. He was packing a bag and I had a feeling that he was about to leave Germany because of the war. I realized that I was up near the ceiling and that I was invisible. Then I fell asleep. When I was in space early on I saw Einstein in space. He said he was busy. We were halfway to Mars I think. For some reason, it was scary to see Einstein in space, but it all lasted about one second. It was around this time that I was in hyperspace. It looked like two colorful planes that were moving at great speed.

When I was in space near the end of my afterlife, I was in deep space and thought that I was somehow avoiding a space war between earth and nearby planets. Perhaps I was dreaming, because God took me from the dream. The next thing I saw was a very large drive-in movie screen in space. It might have been several miles tall and it was playing a Yogi Bear cartoon.

Quite close to my birth while I was still in heaven, I have slipped about two meters to the cloud floor and God asks me if I want to take the shot to become God. I say "no" fearing that it would erase my memories but I want to see what would happen if I had accepted and if God was sincere in his offer. I would become five men lying in a ring and die. This is logical because God is more than the sum of one person. Then I could be in ten places at once, but there would be the risk of permanent death. On top of that, we would get very far from my actual life and there may not be a way back, barring some kind of major catastrophe. Then I was in space, a duality, and I say "no" to the same question again and I feel what it would be like to take the shot and it could cause death. Though I would probably live through it and continue dying from different effects and waking up over and over again in space, perhaps to have an innate understanding of death and to observe the universe or eventually how God created it or he would see how I would create it.

If I had taken the shot in space, there was a chance that I could be born in the distant future aboard a starship. When I think back and remember it, I picture myself walking on deck five of a starship like in the Star Trek movies. God's real intention was to make me into a God like him, but I probably wouldn't last long at first, though I would have repeated chances throughout the process to give up my "life" as a God and return to be born as a man. I would be a vastly different person because the powerful memories of thinking that I was a God in the darkness would erase my memories. The real reason I avoided it was because I was almost born and though it probably would work, it would be too painful to agree to all of this in my right mind.

I was in the Garden of Eden somewhere in Africa hardly alive, and God said that if I wanted to be born in the future to raise my right arm like in a class. I thought I would miss my parents and what would become of them. Also I thought that I would have to look the same and I knew if I raised my arm to make that choice, I would die and maybe still get born in the present, but with no knowledge of the afterlife.

This time Eden was in New Jersey in the future. I had just woken up and was asked to speak by God. I could not speak, but I made some sounds trying to make words this lasted six seconds.

Then I was in Eden in New Jersey and Africa at the same time. I was with Eve and she was standing in front of me and said to leave to be with my starships because the snake would eat me. I woke up and Eden was in the desert. I was in the center of Eden next to a brown tree with no leaves, the snake, about the size of a man, with brown and green spots ate me in one bite. It didn't hurt as much as I thought it would.

Closer to the time of my birth, I had a first person view of God's black jean pants. God said he would make everyone into space manatees. I said "no" that would be bad and gave him words of encouragement and we were saved from becoming space manatees. Maybe this is a joke from God about Boltzmann Brains.

There was a water slide in heaven. There was a rainbow of cloud that was off to the side of the main flat cloud where you could use the rainbow cloud as a waterslide. I was forbidden to use it. I'm sure it was only for children who have a short life, so they can relax and stay awhile. Around this time I awoke in a lower layer of heaven where you could see rolling hills below and a few houses. I saw angels ascending and descending, probably to take care of the dead. It seemed very hot and hard to breathe. I could choose to be an angel, but if I did I would die in two seconds from hyperthermia and exhaustion. Like trying to become a God, I would die rather quickly, be born with no knowledge of the afterlife and it would repeat, each time lasting longer as an angel or God, and I thought to myself "do you want an angel to do work for you that has no knowledge of the afterlife while he is alive?" I think it is ironic, and I'm sure that it has a lot to do with the physics law of the conservation of energy, a universal law where there are only certain amounts of energy in a system, and that energy cannot be created or destroyed. I also saw around this time children playing in front of a small house.

The house was on its own cloud about the size of a small yard, but was connected with cloud filaments to the other clouds of heaven. I think the children there might have died young, and if they did, they might be taking a long break from it because it would be very difficult for God to figure out what their life should be if different from their natural one. Maybe they would go back to it with better luck next time.

One of my favorite memories is on earth. There is nothing around. It is night and I am in a large pink ball. God says that I can marry the ball and the girl will be my wife and we will be gods of some kind. If I choose to say "yes," I risk death and that is just a bad idea. I am not too keen on giving up my life but I'm sure she would be nice.

This time I wake up and see Jesus walking up a hillside by himself at night in ancient Israel. God asks me to say a few things, but I can't think of anything to say because I'm dead. There is a duality here that it is the same memory, but this time I see Jesus walking up the hillside in the same way and God asks me if I'm Jesus, I say "No, I'm Reggie!" Later I see Jesus on the cross in the distance at the edge of Jerusalem and I am asked again if I want to go to the cross. If I do, it might work but I would be trapped in the afterlife. Another duality here is that I see Jesus on the cross from a different angle at a large distance behind and slightly above the cross. The same question is asked of me and I am told that I should do it for the betterment of humankind, but I decline.

I think this memory was created while I was in a pizzeria at the age of about two. I am sitting in heaven and God says to say "Go to hell" to send people to hell. I don't think I said anything and heaven was slightly overcast. There were more clouds in front of me and they were more spherical. It seems that some memories were created when I was young, but I'm probably too old for that now. The memory of the pizza restaurant is real. I really was in a pizza restaurant. But it now seems to me like the event happened in the pizza restaurant, even though I'm absolutely sure that it happened in heaven right before I was born.

This seems kind of quantum to me, to be two things at once. Perhaps consciousness has a quantum nature to it like the position of an electrons probability cloud. Maybe just the formation of memories had a quantum nature because that process might be highly electronic and consciousness is a virtual assembly of a very complex feedback loop of the senses.

Around the time of my birth, I was in a thick cloud again and I was told that this wouldn't hurt. God stuck a golden fork into the top right side of my head. It didn't cut my head because I was mostly holographic, but I it did hurt physically, because it erased information in my brain about what I thought my personality was. I didn't want to remember anything about my life, but I knew who I was as a person and I knew who my parents were. You can't know some of this stuff if you are only a one year old.

When I was in heaven near the end, God said that I had lived to 140 years old. Maybe he was just kidding me, which would be good, because then maybe I can live longer and he will be just kidding me about that, unless I actually live to 140. Then he was telling the truth.

I was standing in front of the cross and there was a gust of wind. Zeus said to God that I was good. Then I disappeared. I saw the manger as Jesus was being born. I was asked to be Jesus and then I disappeared. I remember walking as Jesus when he was a child in the desert near a small oasis. He was with his parents. We were all walking to Egypt. It lasted a few seconds and I woke up and fell asleep several times. Later I awoke in Caesar's camp site as Caesar. I awoke on the cross with two other men. Then I woke up in a flower park in Paris and took a walk around it for five seconds.

Then I woke up flying a bit faster than the wind at low cloud level near the English Channel and God was French. I woke up in a World War II tent for the injured. It was very long and was a dark yellow tent. I was in pain from being blown up. I had to wait about twenty seconds and then a beautiful blond angel eased my pain magically and said I would be OK.

At one point I woke up on a small moon and it seemed like it was the first time I woke up in a similar manner to other events. I was in a very tall mechanoid robot and took a few steps in it on the alien moon. There were bugs on the planet and they were huge, about the size of the robot. It lasted about two seconds. I woke up in space and I was the god of bug planet for one second. They were wearing helmets and had two large radio antennae. The energy between the two large radio antenna held me in place as I felt that I could sense the thoughts and well being of the bug people in a very altruistic manner. It may have been a moon in a parallel dimension. I felt that I regretted leaving the alien moon, but I knew that I couldn't stay for long without causing problems both for myself and them. I felt the presence of God and fell back asleep.

I woke up in a high-rise tower that was very tall, about a mile in height. It was the future and we had a blue "force field" garage door as the side of the building. The apartment had parking for a flying car and was two stories tall with a beautiful view. I was standing on the second floor balcony overlooking the parking spaces below for the flying cars and admiring the beautiful view and clouds out of the large two story "force field" enabled window. Later I woke up in the flying car and there was a problem with the flight electric fields. Our car collided with another due to computer error and heaven crashed. It seemed to me that the collision ended the world of the flying cars. I woke up on the bridge of Star Ship Enterprise as the robot character Data, but I knew I was me. When they said "fire phasers!" heaven crashed and I fell asleep. Heaven said that I couldn't have stayed longer because to kill is bad. Then I was alive!

In space, God said to shake my legs to destroy the universe and I tried to imagine what would do that and then fell asleep.

Heaven was hot, there was a solar flare from the earth and I saw some people. We all thought that it was hot. This lasted about one second. Then heaven was small and was more realistic. Heaven ended while I was in it and it was then that I wanted to stay in heaven forever. At the same time, I realized that heaven's day would get shorter over time and God forbidded it to me. This might be how bugs experience heaven such as butterflies which only live for a few months.

Later I had a dream in my room that I was dancing with a holographic butterfly. It was dancing at a formal event, but was dancing very fast, I realized that it was the female version of God, the one that I saw as the large pink ball in the desert.

At one point in heaven I woke up to a sunset facing the mountain, and God pointed to the sun and it made three passings across the sky. I knew that time was going backwards at an incredible rate. This lasted for two seconds.

Around this time I woke up in heaven and everyone seemed to be asleep in rainclouds. I had a feeling that people were being born. Then I saw God and he asked me if I wanted to be whipped. I said yes. Then I was whipped two times and then I was alive! The whip didn't hurt, it was a magic whip.

I awoke standing next to a wishing well in heaven. If you looked into the well, you could fall from heaven to whatever land that was down there. I saw lots of houses. If I fell into the wishing well, I would die and have to wake up many times just to see my family. I didn't really like the wishing well. I just stood there. God said that he forbade me to use it. I saw other people lounging on the same cloud. The wishing well experience lasted one second.

Near the end of my afterlife, I woke up as George Washington while he was in the midst of taking a footstep in a field near Fairfax Manor. It was a beautiful day. It hurt and I was asked by God if I wanted to be George Washington in my next life. I did think about it for a second, but I decided to decline.

I woke up in space and I played about eight levels of a visual video game with Boltzmann Brain like creatures. I was really good at it, and when I couldn't arrange the colored balls I fell asleep.

I woke up as a cow in a field. It was nighttime and a farmer was standing near a barn in front of his house on the plains. This lasted one second.

I woke up as a higher dimensional object in space. I was a fastly fluctuating higher dimensional box. I wanted to do it forever but it I knew that it was a lot of work that I was doing and that it really wasn't that fun afterall. Again God forbade it to me. This lasted one second. Then I woke up lying on concrete and two men hit me with fireballs like in the video game Street Fighter II. Then I fell back asleep.

I was briefly turned into a locust during the time of Exodus and flew into an Egyptian soldier at high speed. It was nighttime. It lasted one second. The Egyptian was taking a walk near the Jewish area. There were lots of similar houses in a row with a wide road between them and a high wall on the left. The buildings were of ancient design, but they had new awnings that were striped red and white. Next I woke up in a large palace in the middle of the desert. I was the sultan and he had a palace and a harem, right in the middle of the Arabian desert. I wanted so much to be the sultan and then fell asleep.

I woke up inside a giant gold Buddha. It was on the shore of an island and it was as tall as a skyscraper. I could see boats going in and out of a port in the distance. The weather was very nice and the sun was shining. At the time I was in the Buddha, I couldn't see out. Being in the Buddha lasted about seven seconds. Then I was alive!

I awoke in cat heaven and I saw living room furniture. I really wanted to stay in order to have a simple afterlife. Very quickly I had to rise up to higher clouds as I fell asleep.

Chapter 6

Psychology

I might still be in heaven and you might be in heaven right now. The way consciousness and dreams seem to work, heaven and it's workings might be a necessary part of the ether of virtuality. God and heaven might only exist as part of the virtual conjuring of the act of going beyond the physical to become a virtual entity, a mind. That doesn't make it any less powerful, but I can't deny that heaven seems like a place. If it is it is a higher dimension but only in a time related sense, in a spatial sense it is more of a three dimensional place. When we're dreaming there are two dimensional aspects to what we might see, so I would say spaceially, it is somewhere between two or three dimensions depending on the occasion, but it is definitely higher dimensional when it comes to the flow of time. It is like being in the fifth dimension or it has a quantum nature where you can't really say if one thing happened before the other or if it's the other way around.

A hypercube is a five dimensional cube and if you were in one it would be hard to see where the corners were and where the edges were.

Because the universe and multi-verse are infinite, there is an infinite amount of life in it. Even if God is a virtual creation of the mind, some of these virtual creations could be real and the way it is thought up they might work together. This makes God an infinite entity that is real even if his existence is virtual. He would have existed long before our universe in the multi-verse as the virtual creation of other people in parallel universes. He might even create universes because that is a power that we imagine him to have, and added up across the multi-verse he would have that power. I can't imagine any intelligent life anything like us that wouldn't wonder about the universe and not have some ideas about God anywhere on their planet, so it's like the question, if a tree falls in the forest, does it make a sound. If early on in the multi-verse there was no intelligent life, yes there would still be God but it would be whatever animals can imagine it to be. If we go farther, its bugs and then it's microscopic life and God might still exist there, but microscopic life doesn't have an afterlife so does he need to be there? Maybe but his workings would be very physical and you could say that his existence at that point is just the workings of the universe.

Even if some other religion took over and they had opposing views of what God is like, our God would still exist because other viewpoints of it might be rarer. If there is just one planet anywhere in the multi-verse that thinks of God the way we do, then there are infinite parallel universes where it exists there too. This is because of higher dimensions such as the fifth time dimension of destiny which allows for an infinity of possibilities from any action. They would just be more separate but still on a larger scale be infinite in nature.

What we are up against is the brick wall of the limits of human understanding because causality of the multi-verse is fifth dimensional or higher, it's causality might have an infinite or near infinite number of higher time like dimensions. We can only imagine one that flows backwards and forwards like the idea of one dimensionality where the only thing that would exist is a series of points along a straight line. Maybe the sixth time dimension has curves. We sure would like to believe that ours has curves which would signify when we make choices. So again I say that the meaning of life is to make choices, at least that seems important to the existence of life in the universe.

Writing this book has cured me of nightmares. Maybe reading it will cure nightmares. I was very brave but I've had nightmares for decades. It seems that I had nightmares because the flow of time in my dreams was stopped on a very real level. Now I can consider my dreams to be almost good because I can feel the flow of time moving but it is very slow so they aren't good dreams, just OK dreams. If I keep writing this book and working on it, perhaps eventually the flow of time in my dreams will be more normal and I will have good dreams. Perhaps you can only have good dreams and only have nightmares if the flow of time in your dreams is stopped. The philosophy of this book combats the most basic of animalistic fears, that of death, that was useful for eons by animals to be able to fight physically for survival of the fittest, but is no longer necessary in ordinary modern life.

Evil people think that they will have wonderful afterlives even if they sin, but the people they effect go to heaven. If people are ill people in heaven, God will have to do extra work for them. As a result he may punish the evil doers just for this reason, and not give them pleasant afterlives. The same can be said about anyone that does any wrong during their lives, but especially if it effects negatively the present when they die, in heaven or on earth. This is why it's probably good for sinners to repent for their sins especially death row inmates but the terrorists are worse.

Just imagine what a Boltzmann Brain would do if it were caring for people and it knew that the people were sick because of evil doers. It would be furious. The way people are cared for in the afterlife is highly psychological in nature, and punishment from a Boltzmann Brain would not be just physical. It could know everything about a person and if it wanted to, it could inflict great harm on evildoers. Since it may not waste it's time doing so, it will probably just try to get rid of them. Who knows where they go. Probably to a place that would accept them. That place would have to be hell and not nothing because being nothing would probably be too good for them. The realm of true nothingness is God's domain.

They might be nothing for an uncomfortable amount of time and wake up as nothing, which would hurt of course, because even hell does not want to accept them, perhaps until they improve. They will not want to hold onto their mindset of evil if they know they are being denied, because they may be bad but they are not dumb. They will get the idea and then probably have to go to hell.

Heaven's gate is not like a locked door of any kind but is more like the sequence of a movie's production credits at the beginning of a movie, making it impregnable.

Perhaps animals such as fish get a more realistic afterlife than even I did. Their afterlife might actually involve some of the objects that were present in their life, such as for a fish, it's water environment seems necessary. An animals' afterlife might look much more real than a human's afterlife in order to comfort the animal. To keep it out of trouble it may also know that it is dead based on it's heavenly environment.

For humans we need it to look different to know where we are, but the fish is given a more realistic environment. A fish could worry that it was dead but not like a human. The fish is given a realistic habitat that is comfortable for the fish, an underwater environment.

Chapter 7

Messages From The Afterlife

Things we do on some level do contribute to the universe by being actions, perhaps we contribute to the universe by our very existence. Normal dreams seem 40% real because you know it is a dream. Dreams I've had in the afterlife were more real to me as a reality, but I also knew that I was in a dream much like the old adage that life is a dream.

Because God is involved in electricity if he is morphing the universe, then the lightning that strikes people is not caused by his nature. Beyond this, there is a limit to the electrical power under his influence because of underlying physical principles.

God as witnessed in the afterlife need only act where needed and therefore be angels. It is probably as impossible to prove if there is a God as there would be to try to outwit computer programmers of the universe. They could maybe be detected as the lengths of two lasers, which means that the universe may be computed except that it's infinite and therefore perhaps it isn't computed but a random living form. If the randomness of its functional existence could fluctuate as a defense mechanism, then it could be impossible to detect, unless it wants to be detected in an experiment. With the new trend of beings becoming apparent to be prevalent in the universe, because of the detection of so many extrasolar planets in the galaxy. These new discoveries fly in the face of the recurrent historical perspective that earth was the center of the universe. The more recent idea that our solar system was the only one with planets in the universe, which lasted throughout the 20th century. That idea was widely held by most people until after the millennium and telescopes got more sensitive. So why stop there? Perhaps there is be a new form of life that we have known about all along… God, angels, and the physical reality of spirituality itself.

It is possible to have an afterlife and not remember it. It would be an afterlife filled with peace and no pain. This it seems would be forgotten, but even if you had a very active afterlife, you can still get born. It might be as punishment for rising up to heaven or if you just end up standing up and being killed by the lack of an environment. It could be fun like mine, but then you would have to get a more powerful golden fork in the head. The alternative is that you will have knowledge of the afterlife in the next life, but that sounds like a big gift that you may not deserve because it would put everyone in a kind of isolation involving parallel universes that has all aspects the same until your birth. Somehow when you are born, your brain dreams as if you are an adult even when you are in the womb. After getting the golden fork, you would not afterwards think of yourself as an adult, but you could retain knowledge of the afterlife, and by doing so, know the English language at the time of birth.

When I think about why my friend hit me with the mallot and killed me, my only answer is that perhaps I looked different, which kind of spells out the dangers of having a future life as a different person in a very graphic way. This could have very real psychological roots in the person after experiencing death to have a prime desire to get back and to know instinctively that to get back is to be born again. It is mostly a subconscious acknowledgement of the fact that may arise at different times and circumstances in a person's afterlife. The two unescapable realities are, that in the next life, you will be you or you will have knowledge of heaven after experiencing an afterlife and live in a parallel dimension from me. I'm sure that I will have the same life again, which means that as long as the universes can communicate the form of their existence to each other, they would not allow a discrepancy in exactness by having people live in separate universes. We know that the major purpose of the next universe's existence is mostly to be an identical copy to our universe, even if reincarnation is achieved through retro time travel. Especially then, it demonstrates that it is hard to change the destiny of the universe from its natural course. It means you always make the right decisions from your own perspective, and therefore based on equal starting positions for all life forms, we will have the power to choose but we will always make the choice that our soul tells us that we think is right. Since most people have no knowledge of the afterlife when they are born could mean that they did God like activities in the afterlife such as thinking that you stared at the night sky for twenty thousand years in five minutes. I guess you would be a real expert on whatever the subject matter was. This probably is similar to the mechanism that has given me ability to remember photographically events in heaven and the dreamlike existence of the afterlife.

There might be a limit to what you can experience and then get the golden fork in the head, which is explained by the fact that becoming a God or thinking you are one in the afterlife, means that that person will forget the experience at some point shortly after becoming alive in the womb. It could be before, which may mean that you would be more likely to be at peace when born.

You could say that I saved every being in the universe from death by being born because of the anthropic principle. I definitely was meant to be born because I was, and sometimes from my perspective, it seems that I have lived two lives. I can't remember any of its details, except that I was me. So if it did happen, you were there too, but I can't say if it happened, at least not for sure. On the one hand, it would be good because it would prove that we do not die. On the other hand, it may not be necessary because two can equal one.

Because the universe and then life repeats then the afterlife repeats, there is salvation. The extent at which a next identical life and universe can be somehow improved upon in subsequent generations is unknown. Because of a mass theory of information, it may have resistance to change. Another answer to this question is that in the next universe you may live, give or take one plank second, the smallest possible length of time, in either direction. That would mean that if the universe intended to be good, you would live longer on average and that may be a lot of seconds but not that many. To prevent your death, you may only have to live a few seconds longer to avoid the occurrence of a heart attack or the event of a slip and fall. But it may have to be approved in some way through the accounting of the stability of the multi-verse.

Could you create a gateway to heaven? Yes, but it would require the mind becoming a computer's electric field and then you would be alive as the electric field. If it is ever established whether going to heaven entails either an ethereal reality or just one of dreams somehow in between lives but connected fully by being alive both in the present and the past for perhaps a very short time, you could do it. If you were born with no knowledge of the afterlife then your brain will not fit, unless it is in the same exact state where you have no knowledge of the afterlife and think it is your first life. To remember the content of a past life would cause psychological distress, but may represent itself as deja vu. Even though deja vu's existence in the brain involves chemicals, it does not mean that it is not real in some way. Universes being identical indicates that on a deep level, it is the first time you did the thing with respect to all physicality, except your mind. And that is the secret of the universe.

To show that God is science, there are hidden and overt messages in my afterlife that have the same tone of communication and intent that we use when trying to communicate the values of innocentness and good nature of humankind, when we send universal type messages to potential aliens in space such as the innocent and well intended messages sent by the voyage space program by NASA in the 1970's.

It is my belief that we all individually effect the universe in invisible but powerful ways that effect real aspects of destiny on earth and perhaps even elsewhere in creation which we obviously do while we are alive but in equally powerful ways by and during our absence from it. Perhaps the ultimate E.S.P. powers that humans can ever achieve are already being achieved by all to their ultimate power, but in unseen ways. Thereby proving that reality is a collective dream. That it is a compromise between all life forms in creation and the universe, multi-verse itself in an infinite network of collaboration that we call reality.

Chapter 8

GOD on Earth

What I said is that by being born, God performed miracles. Though there is some probability that the birth of every conscious life form is either an equal or a similar miracle called life. The balance in the universe of the act of being alive is of greater consequence to the information structure of the universe than being nothing forever.

Because I was born in America you can say that heaven has declared its footstep on California and Israel. The footstep would be small and only apply to California. This would mean that God loves both Israel and the U.S.A. because we are a free people with just values.

Coming from the clouds on a wispy cloud that had bubbles like sea foam. I could see the town as I flew. I saw a house similar to mine and wanted to be born as a different person. I wanted to be born as every person on earth in some kind of succession. I thought it was necessary because I probably had a lack of faith that other people had been born. God said no and I fell unconscious later to remember this event.

When I was in my twenties I had a spiritual experience at an outdoor antiques fair involving a cloud that I was staring at.

Driving to the outdoor antiques fair, our car passed a long row of tall trees and the shadows of the trees shifted quickly back and forth between lightness and dark. Heaven told me that I was going back and forth between heaven and hell. At the market everything seemed so beautiful and I was thinking deeply about God because he was communicating with me regularly around that time and I was at the time listening to a lot of spiritual type music that was very popular in the 1990's such as Fiona Apple, and Jewel. I was feeling weak from fasting and the long car ride and I thought that the devil spoke to me. I don't remember exactly what he said. I was thinking that he is basically just a different form of God that is the devil himself. I am Jewish so I believe in one God. There is no room for the devil, because God is always more powerful.

There were two areas to the antiques fair, one was up a hill and the other was near us. We bought egg sandwiches from both locations and I don't remember which, but I felt weak from fasting and I thought that in some way one of the food stands represented hell and the other heaven. I was just thinking that I never ate food in heaven and I wanted to improve myself by trying to eat food in heaven because I was feeling spiritual.

When we got back to our showcase I had an attack of anxiety and I thought that when the devil spoke to me earlier, he put an invisible coat of sin on me before sending me to heaven thereby giving heaven a paradox.

As we were leaving the antiques fair I became fixated on a small cloud that was directly above us. I thought that the cloud was God's resting place and that the cloud was "cracked." It was not a dark cloud and there was no thunder. I thought that the shape of the cloud cracked, even though it was probably just moving normally in the wind. I thought that the spirit of God had cracked. I was told that I would be safe if I got into an oil can that was on the field, as an offering like a goat, but I stayed in the car. I knew I would meet God soon on earth and I was worried that being in person on earth would be an ordeal for him. Satan pulled on my seatbelt and I felt strangled. A few days later on the way to the hospital, God's voice was booming from the clouds. The sky suddenly turned dark, seemingly at his command, and the car rode on. I was very sick and I thought that as we drove down in elevation that I was going back in time because God was cracked time had lost its backbone. By the time I got to the hospital, I thought that the hospital and I had fallen in a spiritual way, to the prehistoric time of the dinosaurs. I was greeted by scientists in white coats. I lay down on a bed, and told the man in the white coat next to my bed that I needed to see God. I thought I had done something to God by being involved in the cloud cracking and that therefore I would see God in person. Then I fell asleep. When I awoke, the hospital was mostly empty, and I was greeted by Aaron. He was in charge.

The next day God was standing in the common room waiting for me. At first he thought I was Christian and I said "Are you Jewish?" He said "yes" and he went away and came back wearing a yamaka. He said that I was a good science student at school, and I thought there was something magical in that because how could he know that about me? He also looked just like the God that was in my afterlife and he had the same personality. On two occasions he used what seemed to be magic. The first he said "I can reverse time!" Then I saw yellow holograms float in a circle around him. He said that the man mowing the lawn went backwards. I didn't believe him but the second time I did. A man in his thirties seemed to appear when we were in the outside area. God said to go inside and not to look. As I went inside, I looked back and saw that God was shooting the man with lightning bolts. The man disappeared. I think he was another part of God because God is the sum of more than one person. The younger God that appeared had to be sent back so maybe shooting him with lightning wasn't to hurt him but to transport him. Perhaps he was helping God in a spiritual war somewhere and he came to recharge his powers and go back to the fight. God on another occasion said that he can't kill, and as he said this he caught a yellow butterfly in his hand and let it go.

A real flesh and blood God at the hospital also wanted me to do something for him there just as God in heaven would. I don't know why but it is a similar request to jumping off a cloud in heaven to become God and involves eating a hamburger with ketchup and pretending that I was eating Hitler's brain.

It's that he likes playing around with me but he is very busy and has other things to do, even if he is just a normal person who is doing a lot of activities in heaven and the activities that are done in heaven are magical activities that involve helping people there in both physical and psychological ways. I really can't think of anything that would be fun to do in heaven or that would be purposeful, other than kind of assuming the role of God or perhaps angels to be productive by helping people. Perhaps our idea of an infinite God doesn't really exist. If it does it might be similar in its role to God in heaven by being people who for, whatever reason, get wound up in physical aspects of the universe's existence, or laws, in an effort to help. Perhaps there is a real God who directs these activities, but he too may be a person or otherworldly being who is also wound up in doing work in the afterlife that they have to do to get back to their real life meaning that as Jesus said, God is really not the God of the dead but of the living as he is a living being himself.

It is probably more fun to wish you had magic powers than to actually have them. In this way we are holier than God. God cares about the universe and humanity but he doesn't have to live here. Perhaps this is why he is so fascinated by what we would do as god's. By beings citizen's of the universe perhaps we care about it even more than he does. We would put more sacrifice into protecting it, perhaps in more logical ways than he would, because if God doesn't want to make a certain sacrifice that would interfere with his values of self he can just leave if things don't work out.

At some point in my afterlife I woke up as Jesus in his burial cave wearing the shroud of Turin before he was resurrected. I think this happened while I was at the hospital where I was visited by God but it is not like a dream, and I don't remember it as something that could happen while I was asleep. It probably happened during light sleep where I was mostly awake. In spite of these insights I still can only consider it as part of my afterlife, even though it probably happened when I was eighteen.

I don't know if he was dead or not, but I thought that I was dead though I felt comfortable. Looking at the shroud of Turin while wearing it gave me a powerful sense of a reaffirmation of safety and it still does when I think about remembering it.

To show you just how powerful God is, I was one of my pet parakeets in my third grade classroom before I was born. I can tell that the image of the classroom was messy, as would be expected when looking at the fifth dimension of causality, though it was definitely my third grade classroom. The fine details of that room were mixed with a very colorful cloud of heaven that filled in the blanks. I was one of my pet birds, parakeets. They were in my third grade school room in a cage facing away from the blackboard. For some reason it was painful being the bird. It only lasted for three seconds and faded, but the dream had beautiful colors. This happened at some point during my afterlife, but it is impossible for me to tell which sequence of events it goes with, though it did happen. I think I am not supposed to know, or perhaps because the fifth dimension was used so heavily to make it happen, it has a floaty existence when considering its placement in relation to the other dreams. The fifth time dimension is the superreality of what we consider to be our experience of time. Perhaps it happened near the beginning of the afterlife, which would make God's knowledge of the classroom maybe just be his memory but that is only if I was born before. If not, then he would have had to use the fifth dimension to do it or parallel dimensions where there were similar copies of me. The other alternative is that I have lived forever meaning that the big bang is just an illusion and that the universe is really created externally from its natural function of before and after, as if before our universe existed and there was only the fifth dimension and higher and the universe was created in a top down function originally involving an infinity of higher dimensions that developed eventually into our world and may already be creating at least a huge number of lower or negative dimensions whatever that might be.

There were two times that I saw heaven as the sky. The first was at an antiques convention that was very far from our house. I was feeling spiritual, and when I looked at the sky it looked like the light broke through the clouds in beams of light, as if heaven were near.

This also happened a few years ago. I had done a lot of sewing work and I felt that God was near. When I went outside to take a break, the sky behind my house looked as though the light broke through the clouds in beams of light, and I thought that heaven was near. It was right in front of me, just up in the cloud as a mixture of cloud and light.

Chapter 9

My Childhood

I still have many memories of my childhood, especially those centered around the time of my birth. I remember things about my childhood that at the time, I thought were important memories and for this reason I never forgot them. When looking back at my infancy now, I can only remember what I thought were important memories that I would want to remember later in life but I will never forget my afterlife. I will always remember everything that happened for the weeks duration that I was dead. My memories of it now are the same as when it happened, and when I think of any part of it I can remember what each piece of it looked like in a very photographic way.

It is well known that newly born babies don't understand English and they make baby noises like goo goo ga but I knew English the day I was born. My parents said that I never cried as a baby or made any sounds until I could speak. I never needed to cry because I was happy and I knew where I was.

When I was only several weeks old I was thinking that I wanted to be born in Israel but God said no. I saw Israel in a very colorful vision of a sidewalk in the city and felt like I had visited Israel.

The reason I bring this up is because if you look at my dreams you can see that I awoke and fell asleep many times and felt created by the event horizon of a universe at its end. Also these dreams happened before I was born, and after being born I was thinking that they had happened in my afterlife. On the drive home from the hospital my parents were talking about naming me, and I was thinking about seeing Jesus in the afterlife and I wanted to be named Jesus. Although we know that I told God that my name was Reggie in the afterlife. This shows that I may be a citizen of two universes. The last one that I may have lived in and ours, both identical. It also proves that I knew the English language just days after being born. My first word was H2O, the chemical formula for water. I couldn't form words with my lips but I was thinking them. When I arrived in my room, I felt relieved that I had been born and I was looking at my closet with the light on and felt relaxed. When school was talked about, I didn't want to go. One reason was that I thought I had already gone to school, but I was thinking that without school I would have a problem with math. I enjoyed school and my school was very nice on the side of a mountain in California. A few months into school, I saw the Queen's ship sail across the horizon. I remember the first time I went to a restaurant, but I remember only the pattern of the tablecloth, that I was there, that it was a pizzeria in town and that there was soda.

When I was a baby I was thinking about seeing the flying cars and I told my parents that I was from the future. I really believed that I was from the future. At the age of ten I told my parents that I had a life as a samurai and that I was locked in a pyramid in ancient Egypt in the afterlife.

I have not had a religious upbringing we didn't engage in or talk about religion on a day to day basis. Before getting the golden fork in the head, I was probably as smart as I am now, but afterwards was a fresh start so that things would be new for me and that I would have the identity of being a baby not a twenty something year old man which is maybe the effect of God saying that I was in my prime in the afterlife. I was never an old man in the afterlife. I was in my twenties or thirties. This may be good for all people because then they will be able to identify God by seeing him, though I still hold on to the belief that God may be a person who does work in the afterlife though he was wearing ancient sandals when I saw him in the afterlife and he really did look like the man who I had seen do magic at the hospital and they looked identical.

My great grandfather was a rabbi which meant that he was also the leader of the town. He actually blew the shofar, the ram's horn. That was used to call people to temple to pray.

Personally I'm a Jew for Jesus. It means that I believe in Jesus, but when I need a stronger or more strict mental picture of spirituality, then I can always fall back on God as the solid stone framework of my belief system. I have found that this combination of religions can be extremely satisfying to the soul.

When I was ten years old, I told my science teacher that the universe had infinite parallel dimensions and I told him that Star Trek might exist somewhere as real. This was in 1988 and string theory was not possible, yet he told me that it was string theory but he thought that it was wrong because it had not been accepted by the scientific community and it was not in the news as it is today.

We had a discussion about how I felt that particles couldn't be infinite points and had to be strings, which are ribbons of energy, and probably complicated strings. Ones that do not just have length as their only dimension but width too or even be higher dimensional objects themselves. Perhaps this gives the universe magic powers because on some level it is a Boltzmann Brain, a natural conscious life form that arises from the ever increasing complexities of a system, such as the universe, and could be alive or conscious because it is so complex, just like us. Any increase in complexity of the universe may contribute to how far it's sphere of influence of telekinetic or E.S.P. powers may extend as actuators in the universe. It would have potential "communication" through the web of the multi-verse. Maybe it is very intelligent or complex and can communicate it's information with any parallel universe that it thinks is identical enough to have you born in it.

This does not mean that it has to be alive, only that it can share information with other parallel universes maybe these information networks are complex, enabling the next exact parallel universe to receive you. This would be done as a physical property of the physics of the multi-verse and that there is an ultra exact parallel universe somewhere in the multi-verse, so exact that you will be born there, and being so exact, it will also have me and all of the life in our universe from it's beginning to it's end.

Chapter 10

Dreams of Heaven

When I was in my twenties I went to a Poe concert in Connecticut. The music seemed to me to be as if Poe were a Vulcan from Star Trek like Commander Spock and she was trying to secretly communicate this in her music. Anyways many of the lyrics do involve what can be seen as being attempts at making the album seem like it came from planet Vulcan. Part of one song is "Don't tell the captain, and another is about dolphins, an allusion to Spock's E.S.P. powers, and other lyrics are about black holes and space. I still like Poe and I see this period of my life as a special and definitely very interesting one that I will never forget although eventually I did have to go to a hospital because of the stress I was under to pretend so vividly and for so long, about six months, that I was travelling about the galaxy on starships. The most fun part was when we went to Atlantic City by the Ocean. I had never been there before and it was foreign to me. On the way there we asked a construction worker for directions along the highway, I thought that he looked like God. Maybe I felt lost because God said from his position to me that he would build Atlantic City for me, I guess in a way he really did. Once when we were driving far away from our house, my magic power of pretending that I was in the holodeck started to fade and I was somewhat accidentally telling the holodeck to shut off like they do on the tv show. Since I thought I was on a ship in the holodeck and also on earth shutting down the holodeck from my perspective made the ship crash or more probably could give it strange physics problems on the ship because it had to be on and also off at the same time. When we got out of the car for lunch at a hotel I was having a lot of fun thinking about stuff and I thought that the hotel looked very nice. In a similar manner to the ship breaking I though that the ship had taken me to an antispace version of Vulcan Atlantic City. When we were at the burger king dive through in Atlantic City I talked with opposite God. Opposite God is like our God but he is preconsumed with the negative values of physics compared with our universe and therefore has a different personality and perhaps a shorter temper. They used black lights to check us when we went in the doll convention and I thought they were magic from the lower parallel dimension. I remembered an experience that I had back at the house. That the universe was like an anvil and our universe, the good universe and other good universes were the top spike of the anvil. The bottom spike was basically in hell and involved negative dimensions where people might be bad but physics there was more lax so weird things could happen. It's creator was female and was known by me as the alternate creator and would be something like an atheist Lord, that is, mostly involved in the physics problems of life and it's creation. Our house was several hundred years old and was originally owned by the nearby church.

Taking a walk on the boardwalk by the Ocean from my hotel room to the doll convention I heard God speak, he was maybe embarrassed and trying to help me, he said that the universe was like a cauliflower and that it was his cauliflower and did have different parts. I saw the universe in the sky as a cauliflower and I knew I was right. This was the mid 90's and believing in parallel universes was something of a brave thought, most scientists and science material considered only one universe. Seeing Vulcan New York it was purple and I could see it from fifty miles away because I was so happy that I thought maybe half way that I was on the Vulcan planet. When we arrived in New York, I asked Leonard Nimoy where planet Vulcan was in the universe and he said that it is planet Jupiter. I thought that the Vulcans maybe lived on Jupiter and used powerful force fields to live there or that there was something that we didn't understand about Jupiter. This is interesting because at the Poe concert God told me that there was a secret alliance between Jupiter which was planet Vulcan and earth and that Poe was a Vulcan. This is also interesting because in the Star Trek series planet Vulcan is supposed to be relatively close to Earth and there is a new discovery now in 2016 that there is a hidden planet beyond the orbit of Pluto that is a gas giant and is so far being called planet x. The Vulcan Lord called me Reggie Rog on a few occasions while riding and the library computer also calls me Reggie Rog, perhaps he was telling the future.

When we arrived in New York it was very special for me, I thought that I had made it to Vulcan New York. We went to a restaurant and I remember a woman was standing in front of me and I was sitting, I was taller but she looked so tall from the effects of me thinking that I was in Vulcan New York that she appeared to me as the female Vulcan God and seemed as tall as the sky.

We were on a long car trip in New Hampshire, we went to the mall and I was thinking that it looked like what a rotating space station might look like. Before we were to head home, we went to a local ski slope because I was a recreational snowboarder. When I was getting ready to suit up in the lodge, I was thinking about how hard it would be to manage snowboarding while believing that I was one way or another in a holodeck that had its own physics and this was too much for me. I said that I couldn't snowboard, and the whole trip was for me to go snowboarding. Anyways I agreed to be hospitalized at the mental ward of the hospital there. While the dream that I was on the holodeck was interesting and fun for me for an amazing amount of time, about five months, it was interfering with my life and perhaps my mind ran out of good ideas for making it fun for me and it became scary.

The Eternal Power came to me after I had given up hope at the hospital. He appeared to me as a visual hallucination of a face and it was the only time he has ever done it. He has spoken with me up until a few years ago but he never used God's way of using a hallucination of a face to convey emotion the way God does. I believe God does this because he is more human and emotional and can be deeply concerned about our personal problems. The Eternal Power is an atheist Lord, the most powerful one I know and I did kind of know who he was when he appeared. He is American, highly professional in his seeming rule of the afterlife, and is the person who set up the scene in my afterlife where I jumped into a tiki hut with a suitcase of explosives to blow up a munitions dump as a ghost. He also spoke with me when I was living in Florida in the late 1990's. When he appears, I am vaguely aware that he has a body the size of mine or bigger like ten times the size of a man and is probably usually wearing a suit. I was depressed and wanted to move to a more urban city, and as I looked at a real estate ad, he told me that he would take me to a better place.

From my perspective, I believe he is a humanoid perhaps from a parallel type planet where he lived in the future where deals could be made with their technology for longer and more interesting or fun afterlifes as God's of some kind but he is more atheist of a Lord. He may be involved in some of the physical aspects of the universe or physical problems that could occur in heaven and God would need his help. He described one of these type of problems to me after I got out of the hospital. He said that a very serious problem was to have a conglomerate in heaven where people's minds had gotten mixed up into a conglomerate in the clouds. This may be the type of work he does, and perhaps taking care of atheistic people in the afterlife, I consider him a close friend. If you want to know what he's like, he's exactly like Jack from the Jack in the Box commercials. I didn't know it until around the year 2000 and he is very similar but he is a Lord of the afterlife and perhaps involved in keeping or computing at least some of the laws of physics. When he talks to me I know that he has his own life, and it makes us both sick with schizophrenia, so maybe we have agreed to talk for the last time a few years ago, if it can be avoided. He said he loves people and the universe and that he wouldn't think twice if he had to protect it with his life like The Incredible Hulk.

When I was at the hospital in my room about this time, I saw a higher dimension. It consisted of two floating cities, one was the higher dimensional Hebrew beings and the other was the higher dimensional Japan. They each existed floating on top of giant oval clouds some distance from each other. They were having a war and I saw them launch their most powerful weapons, curses!

Both curses were launched at the same time from their respective cities. The Jewish curse was shaped like a cog that was tube shaped and spun rapidly as it moved. The Japanese curse was a flat black circular plane that had blue balls situated around it's edges and it wobbled as it moved. The curses made contact and the Jewish curse won but I can't be absolutely sure which curse belonged to whom. I was told by a higher dimensional God; she is female, that their entire universe existed inside a light bulb in my room. The whole idea is so cool that I hope that it does really exist!

If it did exist, it would be a good excuse for the proposed escape of gravitons from our universe to higher dimensional space. These higher dimensional planes could float with their cities on top of clouds without the need for planets under them to provide gravity. Gravity would exist everywhere in their universe and maybe at large scales, gravity would seem to differ to them just as at large scales the velocity of the expansion of space seems to differ to us at great distances based on the position in our own universe. Perhaps they think that their gravity differs because of abstract physical laws that they have come up with, but really it is from the positions of planets, stars, and dust in our universe leaking into theirs. This is a new way of thinking about or explaining the laws of nature that really shows that the universe is truly nature at peace.

Our perception and physical laws of the expansion of space might really be just the effect of some kind of energy leaking to our dimension from a lower one where everything is flat and the movements of the probably enormous snake-like beings that live there are what we see here as the expansion and acceleration or deceleration of space and what we consider to be physical laws set in stone. There has been more and more evidence that the physical laws of the universe change slowly over time, making the universe not a stone object, but a living growing form that is life thereby making the ultimate meaning of the multi-verse's existence to be coexistence. It is the expression of life in every possible form across all of infinity in an effort to be something out of nothing.

At the hospital I was standing outside my room in the hallway with the head psychologist. She pointed to the door at the end of the hallway and asked me what was behind it and if I wanted to walk through the door. I assumed that there was chaos behind the door, probably real dinosaurs, and I named it The Nexus Room and I named the lady the Nixit lady because she gave me lots of tests with a flashlight to see if I could look left or right when she asked me to, perhaps to "Nix" some of my thoughts of the holodeck or delusional reality that I had been experiencing for about a year. I wore a straight jacket for an afternoon and I thought that it felt good and made me feel like a responsible man. At one point at one of the hospitals, I said "Let there be light!" because God told me to say it. We thought that they wanted to keep me forever in the hospital. I think that the doctors heard me, and were perhaps amused by it, because I was released a few days later. The hospital with the "Nix it" lady released me after I was feeling better and I asked for pizza saying that the cure to my illness was pizza. Maybe she was amused by this and she bought me a pizza from Domino's Pizza. Probably they released me because in some way it showed that I felt better and all I needed was a pizza and also to them, perhaps that I had my sense of humor back. I was released from that hospital ten days later.

Ten years ago I was up in the desert by myself staying at a house and God projected a fast moving plane directly in my line of vision. It moved from left to right at high speed and had objects in it. I think they may have been fish spirits, but maybe it is a natural working of the brain. It really did look like it might have been a part of the universe's workings though. It seems to me that parts of it were moving so fast from left to right that parts of it could have been moving as fast as I can even perceive movement. If you believe in God, you could say that he was giving me an eye test of a sort; measuring the limit that my mind can visualize movement as a projected plane that moved left to right. It made me feel emotional, I suppose that you could say that it made me think of repentance to life and the real power of God.

Chapter 11

The Spiritual Mechanics of War

When I was in heaven I saw hell from heaven and wanted to be born there in a colorful dream that would have houses and a government, but I would have no knowledge of being born.

Perhaps the only true God is heaven itself and when God appears in heaven it is really just another person that is there that you think is God but really is not.

I can imagine that man made global warming is a fine tuning from God to prevent a perhaps worse dilemma that may have happened in a past universe where he was alive before he became God. The economic toll of using all natural resources on a planet and then not being able to switch fast enough to solar resulting in chaos and a total stop of the world economy as if back to the dark ages could be his "callous" reason for even risking ending the world if we don't dare to switch our energy resources to renewable for our own good.

It might be wise to try to keep violent death row prisoners alive forever to prevent their crimes in the next exact parallel universe because if we kill them, we know they will be born in our next universe and then recommit their crimes on their victims. Perhaps there is psychology powerful enough to avoid them becoming criminals in the next life and it is to control heaven. You could only control heaven from when you first started doing it, because it is a time machine of sorts but if you did and were more powerful or had greater say than the local God in the area of heaven where the deceased death row inmate rests, then you could do magical things to him but it might not work because his reason for being bad is his personality. After all the energy you put in it maybe you could make it work somehow, but you would also have to effect everyone else's afterlife that either knew about it or whose life was effected by it in any way in heaven, with magical rainbows. Then if you could do that to them in heaven, you would be justified in giving the death penalty and knowing that you had not committed his crime by your very killing of him. So it would be best to keep them alive until it can be sorted out in this way, or it may take an uncountable number of near identical universes before the actions of minute changes in space-time fabric with luck would eventually change the lives of the deceased.

Entropy of the next exact parallel universe compared to ours and the energy required for a bullets flight and the loss or gain of one plank second or length or both could eventually result in vastly different outcomes of war over very large periods of the cycles of universal reincarnation, perhaps resulting in more peace over the long term. This would take trillions upon trillions of repeated universes and has to, because it is like the balance of the universe capitulating to exist inside the plank length, the smallest possible length, while also being infinite. Even if the universe isn't alive, the cumulative effects of human wills effect on space-time over multiple generations of the universe might result eventually in a capitulation of peace.

Galaxies with intelligent life might be on some level actual curses between God like beings as some aspect of the existence dilemma or meaning of life, and we should fight for what we think is right because God loves us.

An enlightening break from war. I once imagined a mental hospital that would land on a planet and expand in a circle with windows around the planet at hundreds of miles an hour, giving a breathtaking view.

I think what most people are worried about home robots is that they may break, so there may be special forms of robot insurance in the future with ads on tv. The military is already planning to go robotic with small fighter jet drones. To stop the advance of advanced humanoid robot drones, small laser guided emp bombs, "electromagnetic pulse" bombs that emit a powerful electronic pulse that disables all electronic equipment over an area of many miles could be used. They could be used fired from 3-D printed drones sent by an airship of the future that could have magnetic force fields, nano-warp active force fields, and the power of some level of warp drive or the electromagnetic thrusters being developed by NASA. The combination of which could make a large ship very fast and maneuverable while also being able to carry enormous payloads while being virtually invisible and silent at supersonic speeds. Emp's are being researched as potential weapons of the future but would have to be made small so they only effect small areas.

If you could live in the matrix and whatever you imagined was real, you would use up all of your mental life force on purpose doing what you see as good deeds that in your view both save the universe and help your family. Some of these deeds would be dangerous, such as "time traveling" directly back to one of your favorite memories and being present in the memory as reality of the matrix. This could be bad for you because the computer could have things in the memory turn out differently to make it interesting and enjoyable to you. As your favorite memory, you could have a frustrated or confused response to the memory turning out differently causing pain and eventual death from repeated attempts to perfect your memories or items of the imagination that would lead to a real virtual form of death from exhaustion and pain accumulation, most likely noted by losses to eyesight due to the massive amounts of what is really to the brain and visual system a hallucination that would eventually cause death. If you could live in the matrix, there would have to be limitations to anyone's magic powers. This would also make the matrix more peaceful.

If a nuclear bomb went off, you wouldn't be in pain thinking that the bomb had just gone off. Once God's sure that you have become nothing, just like people think it is, he wakes you up. If you are atheist, perhaps you would stare at the night sky for what would seem like two thousand years and you would be born with no knowledge of heaven, but perhaps your staring at the sky and your ignorance of having had an afterlife, gives you a degree of good luck.

Does God try to avoid horrible things like World War II? My answer is yes, but it means that the event of reincarnation does not absolutely happen totally identically upon each successive life that we live. There can be changes but from what I have experienced I am sure that if it does happen, it is a very slow process that happens only because of quantum fluctuations of time flow by the smallest possible length of time, the plank second at the time of first consciousness or death. Perhaps the moon landing is a spiritual event in some way, and even after World War II has been avoided by the natural workings of the universe and man's will to live over many trillions of lifetimes necessary to jog the destiny of the universe naturally by each plank second per life. At least at first, as the changes were in early development, we would still get born on our birthdays and we would have a similarly or better level of technology when we are born in this distant future where World War II never happens and we still land on the moon in 1969 and have a similar level of technology especially medical technology that some of us do need to get born. On top of this, in spite of there being no World War II our parents would still meet each other although after probably millions or trillions of lifetimes, things could be different but by only changing a very small and extremely highly calculated way in a very careful fashion as if by accident we will be ourselves. There will be no detriment to our lives because of any changes, and those kind of solutions to destiny do require a very powerful and smart God with almost infinite patience. You could say that it is the eventual workings of the natural computation of the cosmos over an almost unimaginable number of cycles of the universe. Perhaps eventually there will be no World War II.

When I was at a restaurant in Los Angeles about eight years ago, I was at a diner and God told me that he had problems somewhere in space that people were at a diner somewhere and they had a war and instead of a nuclear holocaust they had "The Silence Bomb" dropped on them and that he could wake them up. Perhaps this is a symbolic effort on his part either to get me to write my book, because I knew so much valuable information about the afterlife, or given God's somewhat quirky personality that he wanted to drop the silence bomb on me. I think he did in a way when my schizophrenia was first developing in my late teens, so thinking about the silence bomb in LA was a way of affirming that I felt comfortable with God and that we were friends. I am always in touch with him, and it may be schizophrenia, but from my point of view he fits all my expectations of what God should be like and given my history of being born from heaven, I have no reason to think that he isn't really God though I think that he might be the man who I met at the hospital or vice versa because they are so alike. At the very extremes of my belief system it's that he is the man, and when he talks to me, he is the man perhaps living out his afterlife and it is some form of E.S.P. that we collectively experience. When I consider that it is the real God of the universe, I have to consider what that means and give him some slack because there are limits to what God, or what I believe to be God, can show me as fun or to prove that he is God, because even when we just communicate, he is in human form in some way and having limits is part of what it is to be human.

God might be advanced alien beings living virtually; beamed in a certain pattern, tied to space-time in a dreamlike form that can live forever and maybe land on different planets as parts of certain people's imaginations to, in the end, interfere with the destiny of parallel planets to find their own lives. What a waste of energy; or is it?

Chapter 12

Conclusion

On a grand scale it means that the universe is both virtual and real. It might be due to the attention span of the universe. If one part isn't computing a tiny area, strange things happen. What we would consider to be magic is possible due to chaos. It is likely that life is a dream.

You are already partially dreaming when you are awake. That is the nature of thought. If a computer is simulating consciousness and is alive, then the computers chips are programmed to dream it's thoughts. Similarly when you are asleep dreaming you can feel like you are awake in that dream, meaning that consciousness is virtual and also real. We get real inputs to our brain from our senses, but that information becomes virtual when we perceive it because, to be conscious, is to be a virtual conjuring of an almost endless loop of feedback that makes us who we are based on the ways our brains are individually wired. This wiring is both genetic and a product of our environment.

That people perceive the same actual things differently proves that consciousness is a virtual environment. To become a computer, the computer would have to be dreaming about you. This would make the computer simulating you to be only slightly smarter than what it takes to simulate the virtual assembly of consciousness.

In the future you might be able to become a computer by having a procedure done. It would take multiple sessions over a four month period. Different regions of the brain would be connected to computer parts that simulate its functioning. Then you would have a recovery period of one month or until your wounds heal. Then they would put in the next parts and so on. Eventually your brain would be simulated by computer, but if the power died to the circuits you might die. To prevent this, it would have to be self-powering. You would have your five senses. Your virtual brain would also have to on a molecular level experience eating, drinking, etc.

A void in space has been found by astronomers where there is almost nothing except the coldness of space and it measures billions of light years across. The void in space could help solve equations for quantum gravity because it is a simple version of space where the action of laws are more obvious to telescopic and mathematical investigations. If the equations of string theory can be solved, it is said that the advancements to our technology could be rapid and almost unimaginable, like as if we were given technology by a visit from advanced alien time travelers such as the character

and technologies of advanced shows like Dr. Who, the English Sci-fi

television show about a man who can time travel using a device the

size of a telephone booth. This is because string theory is a theory of

all known forces and the equations used to manipulate them, it is

already being used in various disciplines of science and ultimately

could be used to derive solutions to almost any application.

Real type dreams in heaven probably have a degree of actual

reality that is real on earth, at least real in a parallel universe where I

seem to be present. It seems that both during the first and last

quarter of my afterlife, I experienced being in space. This means

that heaven may be a real place and that it is attached to our universe

through the fifth dimension. When I am experiencing being in

heaven and have a lot of real type dreams, I wake up in heaven but I

don't have any emotional feeling that I had just woken up or am not

in the present of what was happening before me. It seems that space

dreams were more comforting because I felt more free and able to do

what I wanted. So the entire part of being in heaven was like feeling

that you had just woken up in the morning but with no pain and

maybe feeling, in the back of the mind, that you are doing a lot of work at least mentally and in semiconscious ways that give a feeling of well-being.

So the meaning of life seems to be to have the freedom to make choices. You may disagree and say that the meaning of life is baseball, but then you made a choice to arrive at that conclusion. The only thing that we do and enjoy so much that we don't make a choice over is love.

There might be more instances or occurrences of love than there are of choices in the multi-verse. People make decisions based on either their love for others or themselves making love to be the meaning of life on earth.

There was a time when I was in heaven that I had an opportunity to be resurrected directly from heaven. I didn't do it but I saw what would happen. It would be very painful and the person would remain in great physical pain with very altered vision from what we consider normal, everything would look like rainbows of color and look fake like holograms. I think that it might be possible to do it in the future with advanced computer and multi-dimension technology to reunite families of people who can't be saved by normal medical means but perhaps they would have to sign up for it.

Also it may be impossible to resurrect people who die before the technology is developed; a constraint of using time travel involving worm holes.

Worm holes would probably have to be used at least to locate the person in heaven and transfer the part of heaven that is simulating them to a robot or reconstructed body back on earth while making their transfer and recovery as painless as possible. Because worm holes are used and that heaven is largely fifth dimensional, time travel would be involved. There might be very great time distortions involved in connecting heaven to earth, and the time jump wormholes only work when you turn them on. One end of the worm hole would be effected by a warp field that points the wormhole to heaven through dimensional manipulation and the other to the land of the living, but the person might have a choice to do it or not, as I did, or I thought I had a choice to do it, but I knew it would hurt.

Our perception and physical laws of the expansion of space might really be just the effect of some kind of energy leaking to our dimension from a lower one where everything is flat and the movements of unseen forces are what we see here as the expansion and acceleration or deceleration of space and what we consider to be our physical laws set in stone. There has been more and more evidence that the physical laws of the universe change slowly over time making the universe not a stone object but a living growing form that is life making the ultimate meaning of the multi-verse's existence to be it's coexistence with other parallel universes in the multi-verse and is the expression of life in every possible form across all of infinity in an effort to be something out of nothing.

The universe is natural but it is an infinite collaboration of the symbiotic effects of an infinite number of universes in a multi-verse that consists of higher-dimensional beings and creatures in the positive and negative two and one dimension with an infinite number of and variation of beings perhaps one of them really is God. There could be an infinite number of dimensions that have only one dimension. There are an infinite number of different kinds of rules that can apply to forms of life that would inhabit one dimensional universes.

The closest thing we have on earth to God is our plants. They are natural and perhaps they have always existed. Plants don't really die or get born, they are the oldest living organisms on our planet and grow because of sunlight and water; perhaps because the laws of physics change over time and universes interact heavily with each other the universe might be alive and therefore God could be real.

At some point the harder it is to directly detect God by science, the more powerful God could be because God is good and the only real foes of God are imaginary from our perspective. They would be beings that exist in some kind of universe. The universe is the domain of life and is therefore under the influence or at least acknowledgement, by the multi-verse network of infinity, which is a part of God.

It's not that the universe is a computer, but that it is digital on certain scales because that is an aspect of string theory that involves part of the reason for the universe to exist and is not more complex or more powerful on an information level than ideas or the reality of God's existence. This shows that the existence of God is more likely than that the universe is just simply a computer.

I think heaven is for reincarnating, but you could get up and do a lot of magical activities like making a soda machine appear and have a coke but then God or the Boltzmann Brain of heaven would have to do a lot of work beyond its normal rest state and would try to get rid of you. This process involves erasing your memory and putting you to sleep. You could be having lots of fun doing magical things with your friends and family but when it's over, God will erase your memory and you will be born on your birthday. When you die, God might arrange your mind not to remember my book when you wake up in the afterlife. This would be, if you go to heaven so you can have some fun or family experiences. Family experiences would be healthy because the real goal of experiencing an afterlife is to have a short break to prepare for reincarnation.

On earth people have lost sight of the purpose of heaven and many think that it is for endless fun forever. That just wouldn't be fun. You would be aware that you had died and not really want to live a whole new life in heaven, knowing that you were dead. You could say that everyone in heaven knows that it's for reincarnation and that they are waiting for the window for them to be born on their birthday and be reunited with their parents, but by being born they forget this and grow to learn that they will go to heaven forever, which is a lie. It's a lie unless you give up your life permanently and become some kind of god, but it would be a hard life filled with tests of fortitude and perhaps just anyone would eventually fail a test or come to a final conclusion of whatever they were doing, like flying across the galaxy, and have to get born on their birthday.

You might think that evolution and new races on Earth are over for modern society, but if people develop E.S.P., it will be a new race and they will be related not by bloodline but by mental intelligence as a race which is similar to a platonic form. There eventually might be a war between the angels where people who control the weather work together to have it be comfortable. Though for people on other parts of the Earth the changed weather could have adverse effects and there would be a war between angels. But eventually it could balance out and be peace. This is similar to today's dilemma of some people having high levels of civilization and comfort in their lives while other people don't in poorer nations. The actual future of people developing E.S.P. powers may not happen evenly and would be a new problem of the future like in the X-Men movies based on the comic book.

Perhaps Jesus Christ did have real magic powers like the ones mentioned in the New Testament and really be from heaven. By having a larger role in perhaps very magical activities in heaven that could involve the teaching of a very powerful part or incarnation of God, he could have very strong connections to it. Therefore the power of heaven would be under his control.

When I think about my relationship to God, I think about him more as if I were his mother because of thinking and remembering being an event horizon at the end of the universe perhaps we are all God's mother making the ultimate meaning of life to be that in an altruistic fashion that we are all here to help each other and even God himself.

Some might say that death is a kind of paradox, although any paradox that can be experienced by the human mind, can be undone with hypnosis and time reversal. I think it is safe to assume that death is the end of all thought. There would not be any thoughts or circumstances after death that could lead to any real paradoxes but what is a more powerful paradox than life itself?

We make our imprint on the universe our whole lives. The universe is changed permanently by our presence and then it is also changed by our absence. Our conscious presence is ripped away from the universe when we die. The universe in general does not give up even the slightest piece of information by way of conservation laws of energy and information.

It would just be illogical to assume that people die and it is an ultimate end. For it to be an ultimate end, as many imagine it to be, would require an infinity of nothingness and would also require what is in essence a period of sleep that would be infinitely long, longer than the expected age of the universe when it comes to it's ultimate end in a big chill, big crunch, or big rip. Since we are not greater in importance than the entire universe, it is inevitable that not any conscious form could stay dead forever.

If the universe were arranged this way, it would take an entire universe for just one person to stay dead "forever." Instead of experiencing the joys of life, they would experience the joys of death. Eventually, because even infinite universes have a beginning and an end, they would wake up and probably think that they were the universe's protector. From then on their lives would consist mostly of the pain of trying to cope while being in some mixture of the opposed states of being either alive or dead, although they too could have their moments of enjoyment of their existence. From our standpoint, they would be gods but being good people, they probably wouldn't admit as much because as the paradox of the parenthood of God goes the only thing that could exist and be what would be from our vantage point a being that lives forever is the cycle of life itself.

Glossary

<u>Actuators</u>-   Apparatuses that move or are effected in such a way as to transfer mechanical energy to other apparatus such as a water wheel that transfers the energy of flowing water to the motion of the wheel that can be used for human purposes such as the Hoover Dam that generates electricity from the flow of water due to gravity.

<u>Altruism</u>-   The want to feel or use the idea of love when interacting with other people or animals. The belief in selfless concern for the well-being of others. The practice of selfless acts done in an effort to help others.

<u>The Anthropic Principle</u>-   The idea that the universe has certain properties of physical law and evolution on purpose because of the obvious fact that the evolution of the universe or multi-verse has resulted in life and specifically higher forms of life such as humanity and beyond.

<u>Apoptosis</u>-   Natural programmed cell death.

<u>Black Hole-</u>   A collapsed star with gravity so strong that not even light can escape. Thought to be connected to other parts of the universe or outlets in parallel universes. Perhaps they are connected through time to the big bang but nobody knows for sure. A very massive star that at the end of its life cycle runs out of the nuclear fuel that holds it up under gravity due to it's enormous heat pressure and collapses but at the same time as it is collapsing an outer shell of the star near the surface around the star suddenly ignites in an enormous explosion so powerful that the star implodes with great force. Forces so great in fact that space itself can't handle it because there are limits to everything and all that remains of the star is a very powerful dreamlike point in space. Instead of becoming just very dense, space has limits, and it becomes just a "point" in space known as a singularity, it's only dimensions being it's mass and spin.

<u>Calabi-Yau Space-</u>   Higher dimensions that exist only on the smallest possible scales. Thought to be the environment of strings (basically ribbon like objects) that dance and vibrate in the tiny higher dimensions thereby being the essence of any particle in the universe. Each particle achieves its identity as an electron or photon

or graviton through the strings vibration pattern and its dance in the Calabi-Yau space thereby making each particle unique only because of its string behavior.

Causality-   The flow of events from one event to the another and the laws involved in its continued existence from one event to another.

Claytronics-   The use of tiny drone like robots that can change shape, color, and arrangements. The ultimate Claytronics technology would be just like the holodeck from Star Trek where in the room different environments can be called upon by a computer.

EM (Electro-Magnetic) Drive-   Propellantless space technology that uses microwaves shot inside a metal cavity that produces thrust from the quantum vacuum energy of space's virtual particles. As long as the space craft has electrical energy it can continue to fire it's thrusters indefinitely.

Event Horizon-   The point where space is warped by black holes and other horizons such as the event horizon of expansion of the universe at very large distances to such a high curvature that not even light can escape it's grip.

The Fine Structure Constant-   A mathematical value involved in the laws governing the interaction of particles and magnetic fields thought to be an unchanging law of the universe.

Goro-   A character from the Mortal Kombat video game who has four arms.

Gravitons-   The theoretical force carrying particles of gravity that would convey the force of gravity between particles of mass.

The Heisenberg Uncertainty Principle-   The property of quantum physics that you can't measure both a particles position and it's velocity at the same time. A way that the universe protects our causality, a property of universal altruistic anthropic values.

The Higgs Field- A field of mass giving particles that extends across the universe. Particles that inhabit the field are called Higgs Bosons which give all massive particles their properties of mass.

Inflation-   It is the big bang or is sometimes considered to have happened shortly before what is considered to be the big bang. It is thought to be a form of infinite power and complexity that eternally creates parallel universes like our own. In fact in the last second that has passed the acts of eternal inflation have just created an infinite number of parallel universes and has done so going back forever in time and will continue to for an infinite of time in the future. In some way it is the creator of the universe.

Manifold-   A region of space that works as a system and behaves from a higher perspective as perhaps an infinitely big flat plane.

Multi-verse-   The whole of creation where all parallel universes exist like the bubbles in a champagne glass.

9.87 meters per second- The natural acceleration due to gravity caused by the mass of earth and the resultant warp on surrounding space that is caused by its mass on the fabric of space-time.

Normal Force- The natural force that keeps a cup on a table when you set it down or for a person standing up and equals the weight of the person or object pointing up.

Plank Length- About 10^{-20} times smaller than the length of a proton, or 0.000000000000000000001 times the width of a proton. A proton, an atomic element, is incredibly small, if an entire atom was the size of the earth, then the proton would only be as big as your house. An atom is so small that if atoms were the size of basketballs, basketballs would be five thousand miles wide. The plank length is so small that because of the indefinite quantum effects of space-time at this very small level, it would be hard to determine where one plank length begins and another ends, because of chaos. If this value fluctuated from one life to another you could eventually avoid the event of your death but this process would take a very long time and many near identical iterations of the universe to have any noticeable effect son your future life.

Plank Second-   The smallest possible length that has any real meaning in our universe and is a quantum functionality of the universe. It is roughly equal to 10^{-43} seconds or 0.0000000000000000000000000000000000000000001 seconds. Ten thousand times a trillion, trillion, trillion times shorter than an actual second on your watch. It is that way because of some very complicated laws of quantum mechanics involving what is known as the plank constant. It is a constant that always arises as the minimal energy increment that can change an electromagnetic wave's frequency, as derived by Albert Einstein. The plank second is so small that in reality because of quantum effects it would be hard to distinguish one plank second from the next or which second happened first, a kind of chaos. In the next life if you live just a few plank seconds longer I don't think anyone really knows how much that it could effect your next life in positive ways. From what I have seen, this level of change is a very slow process, so slow that it would be unnoticeable to you even if you could remember all of the exact events of your previous life as yourself.

Radian- Units of the angular measurement of circles used in trigonometry, the mathematical study of circles and angles. One radian is equal to the distance of the radius of a circle projected along its edge. Pi radians corresponds to the distance along half a circle and is a mathematical law involving all circles known as Pi.

Speed of Light- Objects moving close to the speed of light undergo time travel into the future which is an instance of fine tuning by God to keep light beams on the ship straight and the space travelers alive. The closer to the speed of light you travel the faster is the power of time travel into the future because it takes more time travel to keep the passengers alive.

Virtuality- The property of being real but also a dream at the same time.

<u>Warp Drive</u>-   Being worked on in the laboratory at NASA but on a very tiny scale to test the parameters of the theory of warp drive developed by Alcubierre which was derived from Albert Einstein's equations for space time.

<u>Worm Holes</u>-   Microscopic gateways that connect different parts of the universe and parallel universes through higher dimensional gateways in hyperspace.

www.ingramcontent.com/pod-product-compliance
Lightning Source LLC
Chambersburg PA
CBHW071434180526
45170CB00001B/334